C000056547

Farmer's Girl

Farmer's Girl

ELIZABETH M. HARLAND

Brattleboro, Vermont
STEPHEN DAYE PRESS
1942

*To all
who love and work
on the land*

≥ I ≤

THURSDAY, August the twenty-fourth, was a lovely day . . . one of a succession of lovely days. A pity one had to earn a living within four walls. But perhaps I could get in a walk by the Round Pond on my way home. And anyhow, my desk was near the window, even if the view consisted chiefly of the roof and sooty chimney pots of the cardboard box factory next door.

"Good morning, Miss Harland." Mr Simon, punctual to the minute, as always. "Well, he's got his pact with Russia. There'll be no stopping him now."

He? No need to name him. There was only one he in those days. But was it really serious this time? There'd been so many crises that one was beginning to disregard them . . . to treat them as a joke, almost. To add with a grin whenever one made an appointment, 'Crises permitting, of course'.

"Ten to one we're at war within a fortnight," my employer went on, almost cheerfully. "And about time, too. I'm sick of being forever on the jump, waiting to see what the feller'll do next. High time the boot was on the other leg. About that letter to Jenkins and Jenkins, Miss Harland. You'd better tell them . . ."

The day's work proceeded. But the image evoked by Mr Simon's words continued to dance between the typewritten lines. And though I had my walk by the Round Pond, it was an unsatisfactory affair. Inevitably one's thoughts went back to similar walks in September, less than a year ago, when one had looked at the water, the trees, the gulls, as though seeing them for the first time. When never before had peace seemed so desirable a thing. When, however much one feared that Munich was the mistake it proved to be, one could still delude oneself with the hope that the final catastrophe would be averted, and salve a pricking conscience with an anonymous offering to the Czech Relief Fund.

Now . . .

Mr Simon was right. It had got to stop. Our patience was exhausted, too. Though surely, even now, that didn't necessarily mean war? War . . . all the horror of it, the endless misery, the broken lives, the appalling waste of human and every other material. Surely no man was mad enough to take that last hideous step? Surely, when he knew that we meant business, he would call a halt? He must. When outfaced, the bully always stopped bullying. He must know that appeasement had run its course at last.

Or hadn't it? Didn't he?

The answer, as everyone knows, came eight days later when bombs fell on Poland. And at eleven o'clock on the following Sunday morning we heard the fateful words over the wireless. For the second time within twenty-five years, England was at war with Germany, though on this occasion, the politicians were careful to explain, it was only with Hitler's Germany. On how it was going to be possible to bomb the one without the other, they remained discreetly silent.

Next morning Mr Simon was frankly jubilant.

"What did I say? And this time we'll have to plant the Union Jack on the Reichstag."

Mr Hinton, the cashier, was not so happy. His son, an only child, was in a bomber squadron. Dorothy Playford, the junior typist, came back from lunch with reddened eyelids and a diamond ring, to announce that her boy-friend was enlisting that afternoon. Mary Beeson, the senior, hinted darkly at joining the Army Transport Service. The office boy produced a collection of cigarette cards, and was knowing about makes of aeroplanes. And everyone was being brave about the prospect of being bombed out of bed any night now.

And London went on as usual.

That was really the most incredible part of this strange new world. True, children were evacuated, and people with nothing to do fled to 'safe' areas to do it in. The streets came out in a rash of sandbags. Window-panes were trellised with brown paper or cellophane. Sirens wailed. One listened-in to news bulletins almost every hour, surreptitiously tested one's gas-mask, and endeavoured to develop a sixth sense for dealing with the blackout.

But after those first few feverish days, London settled down to it. The bombers didn't bomb. The evacuees came drifting back again. The ordinary things of life went on very much as though Poland were a free country and no one wanted to hang out his washing on the Siegfried Line.

One breakfasted in the same hasty way at the same hour, ran for the same bus, did the same old work at the same old desk.

Business as usual.

And then, entirely without warning, business . . . as far as the firm of Messrs Simon & Co., importers of this and that, was concerned . . . ceased to function altogether.

One morning, towards the end of September, Mr Simon called the staff together, and made us a little speech, the gist of which was that he'd decided to retire, had sold out to a rival firm, and that our jobs would cease at the end of the week. Everyone would, of course, be given a month's salary. And with the country needing every shoulder to the wheel, he was quite certain that we'd all be happily settled long before the month was out, hrrr . . . hem.

Well, there were innumerable wheels, of course. But the great majority of them required unpaid shoulders only, and shoulders, alas, cannot be detached from bodies which must somehow be clothed and fed. Mr Simon wasn't the only employer who had decided to retire, or had some such decision forced on him. And the 'Situations Vacant' column made about as cheerful reading as the war news.

October was nearing November, and though I was putting in a fair amount of time at first aid and gas lectures, playing darts three nights a week at the A.R.P. centre waiting for ambulance calls that didn't come, and had started on my fourth pair of socks for the troops, I was still officially unemployed.

It was then that I saw an advertisement in the week-old copy of a weekly review passed on to me by a friend.

·WANTED. Intelligent woman, age thirtyish, able to drive car, to help run farm. Experience unnecessary. Particulars from Mrs Charles Rivers, Ashacre Hall, Norfolk.

Madness to answer it, of course. War or no war, I ought still to be thinking about the future. Whatever awaited the 'intelligent woman' at the hands of Mrs Charles Rivers, it wasn't likely to last beyond the 'duration', and the experience would be useless as far as getting another job was concerned. Whereas if I hung on in London something was bound to turn up, and in time, with things as they were now, I might land a really good job.

However, there's something about the possibility (even if it's a million against one) of a bomb dropping on one's head tomorrow, and so settling once and for all the problems inherent in futures, which inclines one to take chances. And I'd never really liked working in an office. Probably I shouldn't get a reply to my letter, anyway.

But I did.

Three days later I received an invitation to spend the coming weekend at Ashacre Hall.

". . . because a dozen letters can't give you any real idea of what the job is like, though I've done my best," wrote Mrs Rivers on the fourth page of her reply. "The only way is to come and see it for yourself. If you'll let me know the time your train is due into Norwich, I'll be on the other side of the barrier, wearing a couple of my best orange chrysanthemums. It's a trifle novelettish, I'm afraid. But it's better than spending half-an-hour searching for each other," and was mine sincerely, Catherine H. Rivers.

I re-read the rest of the 'particulars'.

The farm was Silford Manor, with a thousand acres of land. Some thirty men to oversee (if they weren't called up). Anything up to fifty women in the pea and carrot seasons. Farm accounts, labour books, pay sheets. Herd records for fifty-odd cows. Pedigree pigs. Vegetables to Spitalfields, Borough Market, Covent Garden and Manchester. . . . And I had always imagined, in my ignorance, that farming meant little more than putting corn into the ground and waiting for it to come up!

The salary offered was one pound a week and another pound to 'find yourself'. Lodging, light, heat, and milk, vegetables and eggs in reason, however, were free. If you didn't make a fortune, assuredly you wouldn't starve. And it sounded interesting. It would be fun to watch things grow.

Always supposing I get it, that is, I told myself, as the train rushed through Colchester.

Ipswich . . . Stowmarket . . . Norwich. A crowded platform full of alighting passengers and their attendant friends and relations. Swept along on the tide of them, I passed through the barrier and anxiously but unavailingly scanned the horizon for orange chrysanthemums. The only flowers I could see were the large bunch of Michaelmas daisies in the arms of a disconsolate small girl abandoned by a pile of luggage just beside the barrier.

All at once I sighted a slight dark woman on the edge of the crowd, whose face wore an expression of mingled doubt, resignation and expectancy which could surely mean only one thing — although the pair of dilapidated pink flowers pinned to the lapel of her green tweed coat, while

they might once have been intended to represent chrysanthemums, could never have been orange or merited the description of 'my best'.

"Mrs Rivers?" I ventured.

"Miss Harland? How do you do? Don't look at those
atrocities on my coat. I'd picked two of my prize-winners
for your special benefit just before I started, and then left
them behind on the hall table. And the only florists I passed
on the way in had sold out of chrysanthemums altogether.
The best they could do was orange carnations, which I felt
was most misleading for you. These things were at least
meant to be chrysanthemums, I suppose. So when I caught
sight of them in the waiting-room, I offered to exchange.
. . . Still, they've served their purpose. Amy's outside," she
added, leading the way to the exit, and dropping the offending flowers into a wastepaper receptacle on the way. "We've
got several miles to go yet."

Amy was not, as I had expected her to be, a friend who
had come to assist in passing judgement on my intelligence,
but an elderly, somewhat battered, plum-coloured Austin
Twelve saloon car.

"The farm car," observed Mrs Rivers, as she pressed the
self-starter. "Her brakes aren't too good, and her steering's
a bit peculiar . . . she's rather apt to lurch at lorries, for instance. And if she's feeling sulky, she thinks nothing of oil-
ing up a plug or two and letting you get on with it. But if
you humour her, she'll keep up her end with the best of
them."

And her driver proceeded to demonstrate Amy's abilities
with terrifying skill until the city was left behind, when we

settled down to a comparatively placid progress along a road bordered on one side mostly by fields sloping up to woods; and on the other falling away to meadows through which meandered a long twisting ribbon of river.

"Silford's pretty well off the map," my companion shot at me, as she changed gear on one of the infrequent hills. "No theatres, no pictures, not even a village shop nowadays. The last parson but two used to say that it was a godforsaken hole and the devil wouldn't have it."

"I like the country," I replied meekly.

"So do I. But then I was born to it."

We skimmed over a bridge, ran swiftly through the village I was later to know as Dennington, and took the hill beyond in silence.

"This is where the Manor land begins."

Mrs Rivers was indicating twin fields to our left, a mass of dark feathery leaves stirring gently in the wind.

"Carrots," she vouchsafed abruptly, as she manoeuvred Amy past a large red bus. "Sugar beet," she went on, as we swung off the main road, narrowly missing the overflow from an enormous heap of what looked like parsnips piled high against the hedge. Beyond them the road ran down then uphill between fields of unploughed stubble, each with a long neatly thatched stack in one corner, and more carrots. We turned right, down a short, wide, grass-bordered lane, at the end of which I could see a huge barn, a white iron gate, and beyond it a large red brick house half hidden by firs. I noticed a tractor at work, its brilliant orange paint gay against the dark chocolate brown of the freshly turned earth, whilst behind and above it, silver-white against the grey October sky, gulls wheeled and fought and circled before dropping down to follow in the wake of the plough to search for food.

"How far are you from the sea?" I asked idly.

"Thinking of invasion?" Mrs Rivers' thin face broke into a smile. "You needn't worry. We're a good twenty miles off. And as the old coast farmer said when they wanted to dig trenches on some of his best wheat land, if they do come, they'll take the Norwich road."

"Actually I was thinking of the gulls," I hastened to assure her.

"Oh, the gulls. They're always about in bad weather. I'm afraid we're in for some by the look of that sky. Well, here we are. I'm going to take you in through the yard, if you don't mind, and bed Amy down first. Charles is picking us up on the way back from his shoot."

Avoiding the left turn, which would have taken us through the white gate and up the short drive to the house, we passed through the second gate into the farmyard, ringed about with lichen-walled red-roofed buildings and a large railed-in pond, and came to rest in the garage . . . the partitioned-off end of a cart-shed equipped with two large double doors, and housing at the moment not only Amy but another tractor (not the orange one I had seen on the field, but this time of a pattern reminiscent of a young tank) and a miscellaneous assortment of petrol cans, oil 'drums', spare tubes and tyres of varying sizes, sacks, cupboards, tins of all sorts and a number of tools and odds and ends as to whose use I could not hazard the remotest guess. Whilst overhead in the rafters there seemed to be literally

hundreds of white cardboard baskets . . . a relic of the
Manor's attempt to grow currants, abandoned the previous
season, as I afterwards learned.

"Now for a cup of tea!" exclaimed my hostess, as we
locked Amy in. Whisking me past the riding-stable and on
between a tall holly fence and the ivy-covered wall which
concealed the kitchen garden, she led the way in through a
gateless entrance on to a gravelled drive and through the
open Manor front door before I had time to notice more of
the house than a longish frontage of featureless red brick.
Inside the square tiled hall she paused long enough to call,
"Ready for tea, Mrs Dack," down a passage which I guessed
rightly must lead to the kitchen, and then went straight on
up the wide shallow staircase, at the top of which she turned
left down a landing and ushered me into a small pleasant
room whose one long window was already "blacked-out"
by heavily lined chintz curtains ("Mother had them specially made for the dining room in the last war," observed
Mrs Rivers. "You'll find the others in your bedroom.")
whilst in the grate a cheerful fire was burning, and a gatelegged table in the centre of a terra-cotta carpet was laid
for tea.

"Now I suppose I'd better give you some idea of things,"
said Mrs Rivers, as Mrs Dack, a little greyhaired mouse of
a woman, sidled in with the tea-tray and then crept out
again, closing the door behind her.

The Manor farm, as I had already gathered from her letter, belonged to Mrs Rivers' younger brother who, as a captain in the Territorials, had been called up a week before
war was declared.

"Like everybody else, we'd gone on hoping until the last minute that it wouldn't really happen. But there it was." So Marsham, the headteamsman, had been hurriedly promoted to steward. Major Rivers, an 'Old Contemptible' who had a game leg as a legacy from the last war, and bitterly resented being unable to join in this one, had undertaken to spare time from his own farm to keep his eye on things generally. And his wife, who had been brought up at the Manor and was still 'Miss Kate' to all the men, in spite of her three children and a marriage dating back some fifteen years, had volunteered to act as liaison officer between her husband and Marsham, and cope with the farm accounts (hitherto kept mostly by Captain Irstead's farm pupil Geoffrey Blake, now in the R.A.F.) while she looked round for a suitable substitute.

This latter she had seen herself finding and getting into the way of things within a month at the outside.

"I've got a full time job as a wife and mother and seeing to my fowls," she informed me, "to say nothing of being President and Secretary of our Women's Institute, looking after the District Nurse, and acting as Women's Volunteer Service representative and Billeting Officer for Ashacre and Silford, though so far our evacuees haven't come. Still, we were expecting a hundred and fifty seven of them to arrive at any hour of the day or night for three solid weeks, with half the villagers having bilious attacks through having to eat up all the extra food they'd got in before it went bad. And they may come yet."

But although there had been plenty of applicants . . . members of the Land Army, friends of friends and well

over a dozen replies to the advertisement which was responsible for my being there . . . the post was still unfilled.
Either the suitable candidates had raised objections when
confronted with the actual job, or Mrs Rivers herself had
decided at the end of the interview that they were not
nearly so suitable as they had appeared, for reasons she now
proceeded to relate with a frankness which verged on libel.

One of the snags at which the more likely interviewees
had jibbed, was the fact that they were expected to live more
or less alone. ("Though if the evacuees do come, you'll have
plenty of company, as the Manor is down for sixteen," said
Ashacre and Silford's Billeting Officer thoughtfully.) Most
of the rooms in the house had been shut up, leaving only a
small 'flat' on the first floor, in the sittingroom of which we
now sat. And on the ground floor were the office, the scullery and the kitchen, off which opened the small bed-
sittingroom inhabited by Mrs Dack, who, as the female half
of the 'man and wife' who had run the house for Captain
Irstead in normal times, was prepared to stay on and cook
for the occupant of the 'flat' while her husband accompanied 'the master' as batman.

"Of course you could sleep at the Hall and come up here
every morning, if the worst came to the worst," remarked
Mrs Rivers at this point, her tone leaving no doubt as to the
disfavour with which such a course would be viewed. "But
we're practically three miles away. And to do the job at all
properly, you simply must be on the spot."

I hastened to reassure her.

Since I have always been inclined to be a cat that walks
by itself, the idea of the flat was attractive rather than other-

wise. Assuming that I got the job, I could easily lock myself
into it. And, as Mrs Rivers pointed out, the telephone extension was just beside my bed, though this had been put in
less to calm any possible terrors than because Major Rivers
might have to ring up any morning to change the orders
before the men turned out.

Tea over, we descended to the office, a small room opening off the passage between the hall and the kitchen.

In due course I was to know its contents by heart.

The three enlarged photographs of hunters sailing over
fences at the local point-to-point, with a fourth of Captain
Irstead winning the Members' Race on 'Blue Peter' in '39.
The regiment of files. The framed copy of 'The Fisher-
man's Prayer'. The yellowing newspaper cutting on the
origin of dewponds. The current cropping lists and a sheaf
of Agricultural Wages Board pronouncements. The pair of
ancient prints, the one of 'The Leicester Wether' (a mon-
strous beast the size of a horse) 'bred by Messrs Brown of
Denver', before and after shearing: the other 'respectfully
inscribed by his ob*dt* servant Wm Robinson' to one Mr
Robert Colling of Brampton whose breeding and feeding
had produced an animal more like a barrage balloon than
the seven-year-old 'heifer' it purported to be. An Ordnance
Survey map of Silford, with the Manor land shown in red.
The large, baize-lined, glass-fronted gun cupboard, now
containing nothing but a couple of oily 'pull-throughs' and
a small .410. The shelves above the door, and the bookcase
behind it, bulging with an assortment of articles which included sparking plugs and stationery, a stencilling outfit
for cows' ears, directories and veterinary dictionaries, a di-

minutive pair of shears, insurance cards, a steel measuring tape and a hunting crop. The side table on which is kept the bill-basket, favourite couch of Samuel Pepys, the blue Persian cat. The filing cabinet in one corner. The row of nails from which depend keys of every size and shape, all evidently known by Captain Irstead as intimately as old friends, but which I was quickly driven to label with the names of the separate locks to which they belong (excepting the granary keys which are unmistakable owing to the large block of wood attached to them for the purpose of making it impossible for any user absentmindedly to slip them into his pocket). The wastepaper basket which started life as a pea 'skep'. The blue-and-fawn coconut-matting which in its old age exacts vengeance for its downtrodden youth by perpetually throwing off loose strands for the harassing of hurrying feet.

Now, however, my eye was caught and held by the large flat-topped desk immediately below the shuttered window, whose surface was practically obliterated by ledgers.

"Yes, Nicholas has a passion for accounts," his sister remarked thoughtfully as her eyes followed mine. "I don't think he's let a thing go unrecorded since he first took over. And by simply looking it up in the appropriate book, he can tell you what crop was grown on any given field in any particular year, and the yield per acre. Not to mention the amount of seed and manure sown, and the cost of labour, I shouldn't wonder. Though it's not quite so formidable as it looks, once you're used to it."

We were still involved in account books when there came the sound of a car hooting at the door.

lling
nto
rs
.onacre

... fully occupied at
...uld soon see.

..., about a thousand of them (be-
.. reduction ordered by the Ministry of
..c it had been two) which I was shown next
...ning, the management of the house and the large ram-
bling garden, there were the Rivers children: Alan, aged
eight, George, nearly ten, and Michael, verging on twelve.
Delightful little boys, all three of them, and inheriting their
full share of the maternal energy. And since they attended
a preparatory school in the neighbourhood as daypupils,
riding or cycling the two miles to West Malton every day
and generally returning for their mid-day meal, they were
constantly underfoot at the Hall. Whilst Major Rivers,
though I was to know him as an extremely able farmer and
an astute and independent business man outside his home,
indoors is liable to become merely another demanding
small boy, in spite of his seventy-two inches and ten years'
seniority over his wife.

The evening passed pleasantly enough.

After dinner the children were driven reluctantly to bed,
Major Rivers dozed behind the 'Times', and over our knitting . . . for I had stuffed the current pair of socks into my
suitcase before coming away, and my hostess, saying firmly
"I've made four Air Force helmets, and a scarf for Nicho-

las," had brought out a peach wool vest . . . we
films, books, politics, London, the war, farming an
try life, children in general and the young Rivers in p
lar, our mutual preference for cats as against dogs . .
fact we talked of anything and everything except my su
ability or otherwise for the job for which I had come to apply. Though I had a shrewd idea that behind the barrage of
words Mrs Rivers was busy weighing up my qualifications,
and might already have made up her mind.

I had certainly made up mine. And the long walk over
the Manor fields on which I was taken by Major Rivers next
morning only strengthened my hope that I should be the
successful candidate.

'To plough and sow, to reap and mow, and be a farmer's
boy'. . . . How did the song go? I couldn't remember. But
I longed to try my hand at it, even if my part in the affair
would be chiefly confined to passing on orders for someone
else to do all these things.

At best I had expected to be driven in to catch the 3.15
that afternoon and speeded with nothing more concrete
than the usual non-committal "I'll write and let you know."
Mercifully, however, Mrs Rivers was a woman who believed in direct action. As I was gathering my things together after lunch, she knocked at the door, and entering on
the heels of my 'Come in', went straight to the point.

The job was mine if I wanted it. Could I give her an
answer now? Or would I prefer to think it over and then
drop her a line?

"I don't want you to take it on if you're not fairly sure of
yourself, and then down tools at the end of a couple of

months, so that I have to start hunting round all over again. I want to get things settled, or as near settled as one can in these days," she said frankly.

Could I give her an answer now? Indeed, I could.

"Thank goodness for that! And when can you start?"

I made a few rapid calculations.

"I shall have to collect my things. But I think I could manage Wednesday. Would that do?"

"It couldn't be better," said Mrs Rivers. "I've got to take George in to the dentist's that afternoon. Make it the same train as yesterday, and I'll be on the platform, but without the floral tributes this time."

Downstairs ten minutes later I found the Major in the hall, tapping the barometer.

"The glass is going back," he announced, as one agriculturist to another. "Looks as though we're in for some rain. If we get much more, we shall have to put sugar beet in that field I was talking about this morning, Miss Harland. It'll be too late for wheat."

My career as a farmer's 'girl' had begun.

⊰ II ⊱

IT takes at least five years to make anything of them,"
said Major Rivers, speaking of farm pupils, as we were
walking round the Manor farm the following Sunday
afternoon. "The first year they learn nothing whatever.
The second they notice just a few of the things which they
ought to have seen in the first. The third they learn a little
. . . but not much. The fourth is a bit of an improvement
on the third. And during the fifth they really get going at
last, and learn more in those twelve months than in all the
rest put together. That goes for you, too, Miss Harland," he
added, with a laugh.

I stared at him incredulously. Then, rushing in where
angels fear to tread, I began to argue the point.

It had been decided that I was to spend my first few
nights at the Hall, and I was to move into the 'flat' tonight,
so, properly speaking, of course, the job wouldn't begin until tomorrow morning. But I had crammed a good deal
into the last three days, either working in the office with
Mrs Rivers, or following the Major on his rounds from farm
to farm. And I already knew more or less how my days
were to be filled . . . at least, I thought I did.

Marsham to whistle below my window each morning before the men turned out at seven, in case the overnight or-

ders had to be altered. Breakfast at eight. Round about nine, the arrival of the Major to discuss any necessary business with Marsham whilst I stood by, and to look over the morning's post, which I could then deal with. A walk round, or drive in Amy, if necessary, to see the men at work. Lunch at half-past twelve. An afternoon tour of inspection, to include all livestock twice a week. Back to finish any necessary booking-up and fill in, with Marsham's help, the labour book. Report by telephone to the Major, if he hadn't happened to call in again, and there was anything I (or Marsham) thought that he ought to know. . . . Perhaps there was a little more in it than there sounded, put like that. But not so very much. It was only a question of getting into things, I maintained.

And so it is. Just that.

But the road which has to be travelled in the process is endless, as I was soon to realise. And there are no short cuts.

Your conscious mind may seek to expedite matters by taking copious notes, poring over technical treatises, studying farming weeklies, the Agricultural Station reports, old ledgers and past cropping lists. And no doubt all these help a little. But the real knowledge is absorbed unconsciously, is sucked in as a plant takes up moisture, without knowing how.

To begin with, each farm is a little world of its own. One factory may resemble another factory, one office be very much like the next, but no two farms are alike. Each field is a separate entity, with as many idiosyncrasies as a human being. And in no other occupation are you so much at the mercy of every change in our ever-changing weather. In other businesses it is possible to plan ahead with a reasonable expectation of the said plans being carried out according to schedule. In farming you will often devise a score of plans for a single day, and then have to carry out a twenty-first. And whereas it is possible for a townsman to be successful although himself unable to perform any of the actual operations which take place on his premises, the farmer must not only know what is going on at every corner of his farm, but must be expert practitioner enough in at least a dozen different crafts to be able at any time to detect bad or faulty workmanship.

And this takes not merely time, but a lifetime. A farmer has never finished learning until he gives up farming or 'declines' it, as they say in the farm sale advertisements.

It is only from the outside that farming looks a simple, peaceful and leisurely affair. Once inside the ring fence the tyro is confronted with what at first seems complete chaos. It is only now, after nearly two years, that the pattern is emerging, that I have assimilated enough to reach the stage when my reactions are beginning to become instinctive and automatic, as every farmer's should be. When a change in the weather inevitably begets the thought "Tomorrow we can plough that olland . . . drill the twenty acres . . . get some carrots up . . ." as the case may be. Though a single week was sufficient to knock skyhigh my beautiful visions of farming as something to be picked up in a month or two.

"Confound the people!" exclaimed Major Rivers that first morning, as he went through the pile of letters I had opened and spread out for his inspection. "Look at Prestwick's return!" (Prestwick was one of the Borough vege-

table salesmen.) "A shilling lower than anyone else. And
we're loading him sixty half-bags today. That won't do.
Slip off down to the carrot washer in Amy, and tell Bushell
to work on this list instead, will you? They'll only just have
finished washing the five tons for Manchester. I'll stay here
and run through the cropping list again. I've an idea I've
put down too many peas."

In the carrot house, a huge three-span corrugated-iron
shed built on a small strip of meadow lying conveniently
between the road and a stream on the Ashacre side of the
farm, a man in a yellow rubber apron was busily emptying
sacks of dirty carrots at one end of the washer. A few yards
further on another was removing any that fell below standard as they emerged from a subterranean journey through
jets of water, and passed him on their way to the chute
where empty sacks awaited them. And four more were
bringing up more empty sacks, tying up the full bags,
whisking them on and off a weighing machine, and carrying them out through the lower double doors to swell the
evergrowing pile which later in the day would be collected
and taken through to London by one of Messrs Barrett's
lorries.

Suddenly realizing that I had no idea of which one might
be Bushell, I looked intently at the moving stream of carrots for a moment and then hopefully accosted the man at
the upper end of the washer.

"Major Rivers wants the London carrot list changed for
this," I shouted above the clatter.

Bushell, for I had made a lucky shot, studied it for a moment, whilst I furtively and ineffectively studied Bushell

with a view to being able to recognise him next time. An
indeterminate face, shadowed by an incredibly aged cap
and a large patch of oil, equally indeterminate clothes . . .
the only noticeable thing about him was the yellow rubber
apron, and everyone else in the washer was wearing one of
those.

"Right you are!" Bushell shouted back at this moment.
"And will you tell the Guv'nor that we're nearly out of
Stebbings' sacks? Barrett's lorry didn't bring any in last
Friday."

Manoeuvring Amy past a tractor and trailer on to which
were being loaded the Manchester carrots, which went by
rail and had to be carted to Westham station, five miles distant, I made my way back to the Manor, replaced Amy in
the garage, and returned to the office.

"Leeds . . . he's one of the Spitalfields lot . . . has just
rung up to say he doesn't want any more carrots until the
latter half of the week," Major Rivers greeted me. "The
market's overcrowded, or so he says. Well, he'll have to
have the forty he's down for this morning. They'll be
washed by now. He was first on the list. But we mustn't
send him any tomorrow. In fact, it might be as well to cut
out London altogether for a day or two. What Leeds rings
up today, quite a number of them'll be writing and wiring
tomorrow. Was Marsham in the yard as you came in?"

"I saw him near the cartshed."

"You might ask him if he's any idea how many bags
they'll have on the field tonight. And if it looks like being
anywhere near two hundred, Patfield must tell the women
we won't pick again before Thursday. It doesn't do to get

too far ahead in this mild weather. And we must have
about three hundred left in the house. I must be off. I'm on
the Bench this morning. But I'll probably be round sometime this afternoon. If not, you can give me a ring."

I went into the yard in search of Marsham, but now he
was nowhere to be seen. Lambert, the head cowman, discovered 'washing down' the dairy, was of the opinion that
he might be in the haybarn. A second cowman, as yet un-
identifiable by me, but intercepted on his way through the
yard, half hidden by the forkful of straw he was carrying
over his shoulder, believed that he'd taken some 'pulp' to
the cattle up at Silford Hill. Since the haybarn was only
just round the corner, I tried that first.

Two men, names also unknown, stopped cutting up
wood with a circular saw at my approach, and listened at-
tentively whilst I explained my errand.

"You see which way Billy went, Joe?" enquired the
taller of the two, as I finished.

Joe's reply, given at length, was completely unintelligible.

I had already experienced (and for some time was to continue experiencing) considerable difficulty in understand-
ing the Norfolk speech. It is not so much the words themselves, though several of them were strange to me, as the
emphasis and the tone in which they are uttered. Just as a
wireless set conveys nothing but a jumble of sounds to the
listener unless it is correctly adjusted, so with dialect and
the unaccustomed ear, and it was to take me several weeks
to get properly 'tuned-in'.

"I didn't quite catch what you said," I now murmured
feebly.

Obligingly Joe repeated his information, and this time I
did at least manage to disentangle the words 'Silford Hill',
which I remembered vaguely from my first walk round the
farm was nearly a mile away. The carrot field, on the other
hand, I knew to be nearly half a mile in the opposite direction. I might just as well go there instead, when I should not
only be able to find out the number of bags of carrots, but
also pass on the message to Patfield, who was in charge of
the women pickers. So, turning back into the yard I went
past the pit and the cowhouses and downhill along a track
which ran through a couple of meadows and up a steep
high-fenced lane, before I reached the gate leading to the
carrot field . . . a long sloping field in which I counted
nineteen women bending over large wicker baskets or
'skeps' among the rows of waving green, whilst scattered
over the cleared ground a whole regiment of full sacks
were standing at attention.

Patfield, easily distinguishable since he was the only man
in sight, stopped sorting over a heap of empty sacks to help
me count the full ones.

Seventy-three.

"It's washing day, so we haven't got as many as usual on
the job. But they'll get another hundred, easy, before they
leave off. I doubt the Guv'nor's making a mistake in putting the women off till Thursday, though."

"He's not going to send any more to London till Thursday or Friday."

"Funny thing to me if we aren't in for some rain soon,"
said Patfield, looking up at the cloudless sky. If it was wet
on Thursday we'd be in a nice mess, he elaborated. The

women never liked to work on Friday afternoons because
of their baking. And while a few could occasionally be persuaded to turn up on Saturday mornings, you couldn't
count on them.

"I'd mention Wednesday to the Guv'nor if I was you,"
he concluded.

Back in the office again I booked up and filled in the carrot returns and then hastily consulted my watch and the
notebook in which I had jotted down the various jobs allotted to each man that morning. Half-past eleven . . . and
they all stopped work for an hour at twelve.

Most of them appeared to be 'lifting' sugar beet on 'Gunspear'.

But where and what was Gunspear?

I had seen men at work on sugar beet when walking
round with the Major last Friday morning, but I couldn't
be sure whereabouts. And, anyway, nothing had been said
about the name of the field. Nor had it occurred to me until
that minute that fields had names, though I daresay I ought
to have guessed. Mrs Dack, when applied to, could give me
no assistance. Her province was the house and, except for a
weekly visit to Ashacre church and a monthly one to the
Women's Institute, she rarely went outside it, she informed
me with pride. Marsham, presumably, had not yet come
back from Silford Hill, since he was not to be found in the
yard. The cowmen, who worked different hours from the
other men, had already gone home to dinner. And there
was no trace of Joe and his friend in the haybarn. I decided
that the best thing that I could do was to sit down and to
make myself a map of the farm, and fill in the names of the

fields as soon as I could find anyone to tell me what they were.

One could almost write a village history from its field names.

Some of the Manor's were bestowed in pre-Conquest days. Practically all of them date back several hundred years. 'Doles', though nowadays the mind instinctively associates it with unemployment, is the old East Anglian word for boundary, and across one corner of 'Doles', the Manor field, a long narrow ridge marks the course of a former coach road to London. The faint suggestion of banks across the higher slopes of 'Crabb's Castle' is a reminder that here the Saxons once made a stand against the Danes. 'Leasepit Breck', beside the common, was 'Leech-Pit Breck' once upon a time, and the pond at its western end provided leeches for all the local apothecaries. On the 'Fair Meadow', on every twenty-third of November from 1163 to 1741, Silford held its annual fair. 'Richmond's Place' once belonged to the Earl and Honour of Richmond, reft from its native holder for the benefit of Geoffrey, the Conqueror's son-in-law and, until the beginning of the sixteenth century, costing the village fifteen shillings a year. The 'Three Pightles' were originally three separate 'pightles' or small plots of land. Adjoining the 'Well Piece' is one of the three village wells. The name 'Mautby's Hall' is all that is left of a moated house built by the Mautby family in the days when the Pastons were writing their letters. The 'Monk's Meadow' commemorates a pious gift to the cell of Bec, founded at Ashacre in 1237 and dissolved with other alien priories a century before Henry VIII cast covetous eyes on

monastery treasuries. On the 'Strips', the narrow tongues of
meadow running between the stream and the upper and
lower river by Silford Pools, Silford's one-time paper mill-
ers spread their wares to dry. In the 'Pound Field', behind
the corn barn, once stood the village pound. And as late as
1913 Nicholas Irstead the elder had exercised his rights as
lord of the manor, and impounded a dozen straying ponies
and donkeys, handing over the ransom extracted from their
cursing gypsy owners to his highly delighted men. Almost
every name tells a story, in fact. Although I have never yet
unearthed 'Gunspear', alas! Nor have I learned who were
the unhappy captives responsible for the christening of the
'Upper Prisoners' and the 'Prisoner's Close'.

Nor, to be strictly accurate, did I discover any of the fore-
going on this particular morning, for before my map had
progressed beyond a tentative tracing, Mrs Dack appeared
to say that the butcher had called, and what would I like
ordered for the week? And no sooner had the butcher been
disposed of, than she was back again with the news that
Marsham was at the door.

He'd heard that I was looking for him, he said, when I
joined him on the back door-step. Oh, I'd been down to
Patfield, had I? Marsham beamed approvingly, and I was
suddenly aware that I had passed some private test. And
what about the hoss-corn? He'd mentioned it to Miss Kate
on Saturday. But had she rung up the mill? Do it should
ha' been in by now.

Miss Kate, it transpired, had twice tried to get Dennington mill, but both times the number was engaged. Would I
ring them up now . . . Ashacre 38, the number was down

on that cardboard sheet on the wall behind the telephone
. . . and ask them to send in half-a-ton of bran this afternoon or first thing tomorrow? And how was I getting on?

I told her, concluding with Patfield's opinion of the
weather.

"He seems to think it's going to rain soon."

"Oh, does he, though? He's generally right, too. I'll tell
Charles when he comes in."

The mill was sorry, but owing to the war, they had been
unable to get any bran for days, the mill was at a standstill,
and the lorry drivers at present reduced to digging the
miller's garden. They hoped to have some in towards the
end of the week, however, when they would do their best,
though it wouldn't probably amount to more than a couple
of hundredweight. Meanwhile, should they deliver some
crushed oats to be getting on with? They could manage a
quarter of a ton.

I replaced the receiver and dialed Ashacre 57.

"Crushed oats? Certainly not!" exclaimed the Major,
who answered the 'phone this time. "Not at their present
price. Marsham had better send some of those in the corn
barn down to be crushed when Mark gets back from Westham. About ten coomb, tell him. And keep an eye on them
when they get them home, Miss Harland, will you? Not
even Marsham is to be trusted when it's a question of oats
for horses. If you left them alone, all the horses would be
uncontrollable within a week. What's this Patfield has been
saying about letting the women pick on Wednesday?"

Once again I repeated Patfield's remarks.

"I'm not sure he isn't right. There's been a change in the
last hour or two. It's turning cooler. That'll buck the carrot

trade up, too. Tell him Wednesday, then. We'll decide
about the rest of the week later. Make sure the mill can do
those oats before you send them in."

As I re-dialed Ashacre 38, I became aware of Mrs Dack
hovering by the door. My first course had been ready and
waiting these last twenty minutes. Was it safe to put on the
pancakes yet? Yes, I thought it was, at last. . . . "You can
crush the oats? Thank you." . . . And I took the stairs two
at a time for the flat and lunch.

I had moved the gatelegged table under the window, for
I had already fallen in love with the view . . . to the right
a line of oaks and firs running down towards the island
willow-belt which hid Ashacre's new council houses on the
far side of the river: to the left the seven ancient walnut
trees which shut out the cottages at the top of the adjoining
'Ling Close': and in the centre a vista of meadows reaching
down to the river and beyond, where they merged into trees
now arrayed in all the glory of their russet, gold and reseda
autumn clothes, trees melting in their turn into fields sloping up to yet more trees against the skyline.

But there was no time today to linger over views, however lovely. It was already one o'clock. And if I failed to
catch Marsham in the yard and tell him about the oats
within the next five or ten minutes, there was no knowing
when I should find him. Then there was Patfield to revisit,
so that the women could be warned about Wednesday before they left off. And what was to be done about my tour
of inspection? What was happening to Gunspear and the
sugar beet? It looked as though I should have to fall back
on Amy.

The Manor land, unlike that at Ashacre Hall, which lat-

ter forms a neat compact rectangle, sprawls clear across
Silford (itself a long straggling parish) from one boundary
to another, a good two miles from end to end, and over a
mile in width at its widest point. Gunspear, as Patfield helpfully told me when we had settled the question of next
picking day, was up beyond Silford Hill, something like a
couple of miles from the carrot field in which I now stood.
But it was such a heavenly day, the sun high up in a sky
still cloudless in spite of Patfield's prophecy (which he had
repeated twice within the last five minutes): and I could
detect nothing of the lowered temperature mentioned by
Major Rivers . . . the air was more kin to summer than
late autumn verging on winter, it seemed to me . . . so I
decided to leave Amy in her garage, and struck off across
the fields in the direction indicated by Patfield, stopping
now and then to make what I hoped were useful entries in
my notebook as I went.

An orange tractor, a 'Fordson', I note carefully, towing
behind it a machine which seems to be scratching up the
stubble, is 'riffling'. I discover this because the driver, a
fresh-faced blue-eyed shortish man with a most cheerful
smile, who might be any age from twenty-five to forty
(thirty-nine I learn long afterwards) pulls up at my approach, imagining that I want to speak to him. But he has
restarted and is halfway to the distant fence before I think
of asking him either his name or that of the field, so for the
time being both remain anonymous.

The next field is full of carrots. But why are those halved
motor tires lying about in odd corners? "Water for the
young birds," the Major informs me later. "Even the silli-
est young pheasant can't drown itself in one of them."

In the lane beyond three men are trimming hedges in a leisurely manner which quickens considerably the minute they catch sight of me. Over another field of carrots, across the Ashacre road, and I follow the 'Sandy Loke' up to Silford Hill, where I have been told to look at the 'store' cattle two or three times a week and have decided on Mondays as one of the times but they will have to wait until tomorrow. And so to Gunspear and the sugar beet.

"How's the sugar coming out?" I am asked by another unknown (I shall be driven to suggest that they wear labels for the next month or so) and have to confess my ignorance, though I have a vague recollection of being shown some little yellow slips last week by Mrs Rivers.

"That didn't ought to be too bad t'year," vouchsafes a tallish loosely made quiet-voiced individual whom later I am to know as Will Long. "You can always tell when there's a decent lot a' sugar. The beet turn black after you ha' topped 'em."

He picks up one of the beet, and I rub my finger across its moist darkened surface, then lick the finger clean. It certainly tastes sweet. I also learn that the beet must be well 'topped' before carting, or the factory will deduct a considerable weight from the grower's return. And that the tops themselves are a valuable food for cows. To my unfortunate inquiry as to how many weeks it will take to finish the field, I receive the slightly amused reply that they hope to start on Broomhills . . . Broomhills? I must get on with my map! . . . tomorrow afternoon.

On my way back to the Manor I am overtaken by the Major, who pulls up with a screech of brakes, to say that I can save him a journey if I'll tell Marsham that he's changed his mind about ploughing Coney Hill next. Vincent (my blue-eyed tractor driver?) had better get on with the Easter Field first, and leave Coney Hill until afterwards. We were certainly in for a change in the weather, and if there was much rain Easter Field would be too sticky to touch for weeks, and we'd probably never get it ploughed before Christmas at all.

"And you'd be surprised what a difference autumn ploughing makes to the crop."

Then I'd better ring up Barrett's, the haulage people, when I get back, and ask when they are taking the next lot of beet into the factory. That heap at the top of the Ashacre road has been there quite long enough. And warn them no London carrots tomorrow. If they'll ring up about midday we could tell them about Wednesday then.

I reach the house to find Mrs Dack twittering about in the hall. What time would I like my tea? The kettle had been boiling quite a while, as she thought I'd have been in by four.

A cup of tea! And I suddenly realise how glad I should be to sit down for five minutes in the flat. I could have my tea first and then ring up Barrett's from the extension.

But I have barely started for the stairs when Marsham appears to collect the advice notes for the London carrots. And when would I like to do the labour book?

The advice notes are ready on the office desk, but I have forgotten all about the labour book. Nor have I reckoned on Marsham's friendly interest as to 'how I'm a' goin' on', which makes the booking up of the said labour a lengthy business. Whilst my message from Major Rivers leads to a somewhat one-sided discussion on varying soils, 'light' and 'heavy' land, and so forth, which I do my best to follow with intelligent interpolations, though a good deal of the doubtless valuable information I am given is wasted, owing to my inability to grasp more than about a third of what is being said to me.

My tea, when at last I pour it out, is abominably stewed, for Mrs Dack has evidently taken it upstairs when Marsham first came in. And having finished the labour book, it has occurred to me that I might just as well ring up Barrett's from the office and be done with it. And so I have.

Still, I ought to be able to call it a day now, I tell myself. And decide to wait and hear the six o'clock news, which is almost due, and then have my bath.

"Please, Miss, Peachey says can he have some paraffin oil for his hurricane lamp?"

Lamp paraffin, as distinct from tractor paraffin, is kept locked up in the 'copper hole', an outhouse beside the game larder. I have the only key, and have been warned by the Major that, as with the multifarious 'stores' in the outhouse in the pumpyard, unless I issue it personally, I shall never know where it goes or when it is running short. So I descend again and accompany a shadowy Peachey, armed with a torch tied up in a spotted handkerchief and an oilcan, across the courtyard outside the back door. A pity I can't see more of him, then perhaps I might know him again next time we meet. Marsham and the labour book between them have told me that he is one of the teamsmen, and that at the moment most of his time is spent on carrots, ploughing them out before the women pick them, carting them down to the washer in a large wagon drawn by two, sometimes three, horses, and helping Barrett's men load up at night. But so far I have never come across him at work. And tonight he is little more than a soft rather high pitched voice.

"Nice business, all this here," he vouchsafes, as the paraffin runs out of a fifty gallon drum into his can. "Can't see your hand in front a' your face, dussent show a light indoors or out, in case a bomb drops on you, prices goin' up all over the shop, and all because of one man."

"Hardly that," I demur, as Peachey turns the tap off. "You don't suppose that if you tried to be a Hitler in Silford you could do anything unless other people were prepared to follow you, do you?"

There might be something in that, Peachey concedes after much thought. He can't say as he's anything against the Germans theirselves. He doesn't happen to have fought agin them in the last war, not having bin called up. But there was a batch on 'em in that old tannery on Banwell Common for nigh on a year, what the Old Guv'nor had for a month mudding out the river. They'd seemed a decent enough lot a' chaps, for all one on 'em got fed up one arternoon, and hulled his crome into the water. Still, there 't is.

"It is," I agree sadly, and lock the door behind him.

It is too late for the six o'clock news when I get back to the flat, but I can always listen-in at nine. The bath water is boiling, and I soak and soak in it. Then I dress, write two letters, and Mrs Dack brings in my evening meal, followed by Master Samuel Pepys, who dislikes scrambled egg, but sits on the arm of a nearby chair reproachfully eyeing every

mouthful I eat, and finally condescends to accept a saucer
of milk.

The remains of my meal cleared away, I pick up a book.

One of the great advantages of my solitary evenings, I
had told myself, would be that I could get in quite a lot of
reading. And I had had a thoroughly enjoyable burst of
extravagance before leaving London on the strength of it.

But the words do nothing but blur. I read the same sentence three time in succession, and am still wondering what
it is all about. Instead of listening to the news, nine o'clock
finds me in bed and fast asleep.

Patfield was right about the weather.

Two mornings later, as I lay awake listening to the noises
which were to become a familiar prelude to Marsham's
whistle — milkmen shouting to the cows, a series of dull
booming thuds (Is it bombs in the distance? No. It's only
Lambert rattling the churns this time), the horses surging
out from the horseyard to drink at the troughs beside the
pond, the protesting quacks from the ducks their approach
awakened — I grew gradually aware of a soft swishing
sound. When, half an hour later, Marsham arrived under
my window, it was to announce that we had had twenty-
eight points of rain during the night.

Hitherto rain had been merely rain to me. Something for
which I put on a mackintosh or put up an umbrella. A
nuisance, to be sure, but usually resulting in no more serious
dislocation of my day than damp clothes, or mud-splashed
stockings to be washed when I got home. Now I was to meet
it on an entirely different footing. On a farm it is a power

to be reckoned with, its absence likely to cause as much disaster as its too copious presence, and even a shower of it apt to involve innumerable changes of plans and orders. Whilst at the Manor its visitations are measured in a rain-gauge, recorded in a note book, and totals added month by month to a chart on the office wall already summarizing the rainfall of over a dozen years.

Twenty-eight points, I guessed from Marsham's tone, was a fairish amount (actually it is over a quarter of an inch) but since it had followed a series of dry windy days and was now lapsing into a fine drizzle, it ought not to make any alteration in this morning's orders, bar stopping Vincent, who wouldn't be able to do anything on Easter Field until it dried off.

"Still, you'd better just ring up the Guv'nor and ask what he thinks about the wheat drilling."

The Guv'nor is always Major Rivers, just as the 'Old Guv'nor' means Captain Irstead's father. Captain Irstead himself is always referred to as 'The Master' or 'Mr Nicholas', a distinction whose subtlety, if subtlety there is, I have yet to fathom.

"It shouldn't make any difference to that light land on 'Gypsies'," replied the Major, when roused. "But if he's at all doubtful, tell Marsham to run up and have a look, and if necessary, let Roger get on with something else until nine o'clock."

By breakfast time the last lingering drops had ceased to fall, and a keenish wind was at work chasing stray leaves from the trees. 'Gypsies' had taken no harm, as I found on visiting Marsham's son Roger later in the morning. The

wheat was going in a treat, he told me. And when I entered
up the labour book at the end of the day, I was able to write
'finished drilling' and the date against the 'Gypsies' field on
the cropping list.

Next day, however, was misty and dull. No wind came to
scatter the lowering clouds. Moisture coursed down the
Manor walls indoors, and outside trees dripped, bushes
when touched sent up a fine film of spray, and tiny beads
of water spangled my hair and the men's moustaches.

On Friday morning another thirty points of rain fell between ten and twelve, sending all but the most hardy of the
carrot pickers flying home on their bicycles. (And London
had already begun to wire "Load heavily.") Saturday
brought a deluge, a hundred and thirty-six points, and so
much of it before midday that most of the men left off early
and went home draped in sacks already sodden. ("Have to
watch an' see they bring 'em back, too," confided Marsham
as we did the labour book. "Do we shan't see one on 'em
agen, and we've lost enow as it is, what with blackouts and
such.") By Sunday afternoon the river had overflowed its
banks and inundated the meadows below Silford Bridge.
Lakes covered the greater part of those which made up my
view. And in the hollow on the Ashacre road which lies
midway between the top of the lane leading from the
Manor and the bottom of the 'Sandy Loke', water a foot
deep covered the road for a distance of nearly sixty yards.

Attempting to ford it with Amy the following afternoon
on my way into Meddenham, a small market townlet four
miles distant, to which I had been sent in search of ploughshares, I succeeded in splashing safely through. But on the

return journey the engine coughed, spluttered and died when there were still ten or a dozen yards to go, and I had to be rescued eventually with the help of two of the farm horses.

"Can't anything be done about it? Aren't there any roadmen?" I asked Marsham, as we pushed Amy into the garage.

"Well, you see, it's like this here."

The hollow had always been a nuisance whenever there was a heavy rain, it appeared. But as it didn't happen as often as all that, and the Old Guv'nor, though he'd grumbled, hadn't seemed to mind very much, the roadmen had got into the habit of letting the water through on to the adjoining Barn Breck. For the last few years the water had done no more than form an outsize puddle which, though impassable by un-rubber-booted pedestrians, could be successfully negotiated with cars and bicycles, and had usually seeped away within forty-eight hours. Now look at it. And all the Council did was to hang up a couple of red lamps at night, said Marsham disgustedly. Two baby Austins, as well as Amy, stuck in it already. And I should have seen the mess Isbell got hisself in, coasting down the hill on his bike the night afore, and off in the water afore he knew it was there. He didn't half swear, neither. And it 'ud be there for the best part of a week if we didn't let it off on to the Barn Breck.

I rang up Ashacre Hall.

"I know. I was just going to give you a ring," said Mrs Rivers grimly. "We've only just managed to scrape through ourselves. And I've got to go into Meddenham tomorrow. Nicholas would swim it first, I daresay. But he's got something else to think about in France. So you can tell Marsham from me that he can take a mattock to that fence right away. Amy? Croft must take her plugs out and dry them by the kitchen fire. If she won't start in the morning, ring up Paynton's garage . . . Meddenham 68. . . . They do all our repairs."

Croft: groom, gardener, and handyman, but no engineer, failed to elicit a single spark from a thoroughly disgruntled Amy next day. Paynton's man, when at last he came, announced that water had got into the batteries, that they'd have to be recharged, that he was afraid they hadn't got any spares just now to lend out meanwhile, but they'd do their best to see that Amy's own were returned the day after tomorrow.

Still the rain continued, bent on making a record for November (which it achieved). The half-dozen delphiniums, the single lupin and the few monthly roses left upholding summer's standard in the kitchen garden, hauled down their battered flags. The land, as Marsham remarked at least once a day, was like a sponge. Wheat sowing was abandoned until the spring, with only sixty-three acres in, and a war on at that. The stream of carrots to London and Manchester dwindled to a mere trickle. The sugar beet lifters, on piece-work, and accustomed to drawing quite a bit extra on pay-day, could hardly earn a week's wage. And plans for the day's work were continually revised or abandoned. Long before the month was out I had acquired a new and wholesome respect for rain.

I had also made more than a bowing acquaintance with the cow records.

"See that Lambert brings them in to you once a week,

after weighing day, with a statement of what milk has been sold to the men, how much has been used for the calves and in the house, and how many gallons to the Milk Marketing Board, so that you can check up on it. Now that they collect only once a day because of the petrol shortage, it's quite simple. Then you look over the sheet to see if any cow's yield has dropped suddenly, and if it has, ask why. But as each sheet covers four weeks, you needn't book it up in the record sheet oftener than that," Mrs Rivers had said, when explaining the system. And added, "And when you get this month's in, there'll be all the yearly totals to get out, as our recording year ends on October the eleventh. Still, another week or two won't do any harm, so you might as well save it for a really wet day."

At various times I had put in a few odd minutes on the records. Now I was able to finish them and make out a species of examination results list, with the cows graded according to their yearly yields, and a line drawn at seven hundred gallons, below which came the 'failed' and 'doubtfuls'.

"Now you'll have to get Lambert in to go over it with you," Mrs Rivers, who had dropped in to tea, said to me next day when I showed the list to her. "And when you've done that, get Charles to have a look round them. You'll find that several of those who look all right on paper will probably turn out to have something the matter with them when you see them in the flesh. And a few of the not-so-goods may have some excuse for it and be worth another chance. Though I don't believe in being too lenient. Nicholas isn't really interested in cows like Father was. And he's left far too much for Lambert, who's got a bit slack, in my opinion. When I used to keep the records, before I married Charles, we had far more over the thousand gallon mark than you've got there. And your herd average is a good two hundred gallons below ours. It was nearly as bad last year, too. You'll have to keep your weather eye on Master Lambert."

In the office next day Lambert, giving off a fine aroma of mingled cow, cowhouse, and disinfectant, studied my list dispassionately for a moment or two, and then began to displace several successful candidates.

"There's old Duchess. She's not in calf, nor won't be agen, neither. She ought to be fatted up, by rights, as soon as she's dried off," he commenced.

I pencilled a disparaging note beside the ducal lady.

"And Cherry. It's going on eighteen months since she had one, and she ain't in calf now. She won't milk much longer. It's no good keeping her on."

I made another note against the unfruitful Cherry.

"Sunday, she slipped agen last month. That's the third time. And there's Marina an' all. And two a' them Irish heifers what Mr Nicholas bought two year ago. You might ask the Guv'nor if we can't have another old billy goat. The Old Guv'nor allus had one running along with the cows. And there warn't none a' this trouble then."

I recommend this remedy for contagious abortion, long practised in Norfolk, and whose efficacy is still firmly believed in, for the attention of the Royal College of Veterinary Surgeons.

Marleen was a homebred, and so wild in consequence

. . . the Irish and such got so much handling and knocking about before they came into the herd that they were
rarely any trouble Lambert confided wistfully . . . that she
still had to be roped before anyone could milk her. She'd
been a long time dry last year, too. But she was beginning
to get into shape now, he shouldn't wonder, and he thought
she'd pay for keeping another year, to see how she went on.
Aldershot's first calf had been born dead on the Duck Hut
meadow when nobody was expecting it for another month
at least, and it had been a day or two before he'd come across
her. So Mr Nicholas had said to give her another chance.
Dolly was due to calve in four months time, so it was no
use getting rid of her yet. And she might come in a bit better this time. But Roanie had started 'wasting'. It would be
money thrown away to try and fat her. And if we didn't
get rid of her soon, the vet would likely as not order her to
be destroyed next time he inspected the herd. He didn't
know what to say about Catherine. Would I come and have
a look at her?

Unable to believe that any look of mine would be of the
slightest value, but feeling it my duty to endeavour to learn,
I meekly followed him into the cowhouse where Catherine,
a bulky red cow, stopped eating sugar beet pulp for a second and turned on us an unexpectedly white face with the
eyes red-ringed like a clown's mask.

"What's wrong with her?" I asked, unable to detect anything obvious.

"Garget," returned Lambert in a sepulchral voice.

"Garget?" I repeated blankly.

"Mastitis, the vet calls it," supplemented Lambert.

I remembered a pamphlet headed 'Mastitis in Dairy

Cows', which I had seen sticking out of the office letter rack, and determined to acquaint myself with its contents forthwith. Meanwhile, what was being done about it? I asked.

"If that Inspector chap sees her like that, he'll order her out of the herd," prophesied Lambert gloomily, ignoring my question.

"Surely it isn't as serious as all that?" To my ignorant gaze Catherine looked aggressively healthy.

"Well, I dunno. She's had a touch of it afore, and it's bin gone afore he came round. But it's two quarters this time."

Major Rivers was due at any moment, so I retired into the office to look up the pamphlet and was halfway through it when I heard his car outside the door.

"That's Lambert all over. If I'd been Nicholas I'd have told him to break fresh ground years ago," remarked the Major cryptically when I laid Catherine's case before him. "Of course Johnson will have to report her if it doesn't clear up. But nine times out of ten it's only a temporary condition after calving. Or she may have got a knock. Her milk mustn't be mixed with the rest, of course. But it can be used for the calves. Tell Lambert he's not to turn her out during the day, but to keep her in the cowhouse, wash the udder well and use plenty of 'relief', and strip her four or five times a day. Half-a-minute, I'll come into the cowhouse with you and tell him myself."

So the clown-faced Catherine was reprieved, though I was a little disconcerted to find that my zeal earned me anything but compliments from Lambert, as Marsham informed me with relish when he came in to do the labour book.

"You don't want to take any notice a' what he says,

though," he went on, with a chuckle. "You should hear how he goes on about me whenever I has to say something to him. Harry allus likes to do all the fault-finding hisself. He always starts mobbing and blazing away the minute anybody looks at him. And there's plenty to look at, what's more. I don't say nothing against a pint or two, meself. But I never could see the sense in too much on it. I take arter me feyther. A pint every Saturday night was what he went in for. Allus reckoned as it was as good as a dose of med'cine to him. So it was an' all. He never had more'n the one. An' he stopped that arter the way he was done down by old Joe Taylor, what used to keep the 'Hart' in them days.

"Half-a-thick'un and a sixpence he had in his pocket when he went in. I've heard him tell the tale many's the time. Beer was tuppence a pint then, and he handed over what he took to be the sixpence. And Joe, he gav' him fowerpence change. It wasn't until he called in at the shop on his way back home that he found as he'd still got the sixpence in his pocket. Back he goes to Joe at once. But Joe, he wouldn't have it at all. Faced him out that he'd never given him more'n sixpence in the fust place. 'Right you are!' says me feyther, an' swore he'd never set foot in the 'Hart' agen. No more he never did.

"Well, I'll be sayin' goodnight, miss, and getting along home to my tea. And don't you pay no regard to Harry. You're a' doin'."

❧ III ❧

THE Manor week ends on Thursday night, and the men are paid on Fridays . . . a custom thoughtfully inaugurated by the 'Old Guv'nor' to enable the wives of his employees to shop on Saturdays.

In theory the paybook is a simple affair: a mere consulting (if you don't already know them off by heart) of the current rates of pay as set out on the latest Agricultural Wages Act sheet, and the deducting of insurance contributions. In practice, however, it is a highly complicated business, since scarcely a man on the place ever draws the exact statutory wage. In the course of years odd bonuses have grown attached to this and that. Small rises or special overtime rates have been awarded to particular men. Peachey, the second teamsman, for instance, who carts most of the carrots from field to washer, draws an extra sum weekly as long as the carrot season lasts. So does Joe Gidney, as Mark's mate on the road work tractor and trailer. Mark himself is entitled to driving-money for every hour on the road, and his skill in mechanical repairs has long ago lifted him above the ruck of ordinary wages. Peel, who numbers driving the threshing engine among his accomplishments, and is also useful in the 'fitter's shop', has a wage slightly under Mark's,

and draws 'board money' for every day's threshing. So does
the 'drumfeeder' (Kettering, as a rule): and if the corn
threshed exceeds a certain number of coombs, every man
on the stack gets an extra shilling or so. The carrot washers
receive an extra penny for every hour spent washing carrots. A penny a bag is paid out over and above the hour's
rate for all carrots picked on the field. Drilling corn and
the sowing of soot or artificial manure entitle driller and
sower to so much per acre, per bag, or per ton. Isbell, when
cleaning out the meadow streams, is sometimes paid so
much per chain and at others ordinary wages plus a small
additional sum for each day of river work. Peacock, when
carpentering, gets an extra sixpence a day. Patfield receives
a percentage when empty 'pulp' or manure bags are sold,
and a bonus reckoned on the total crop when peas and carrots are over. And so on and so on. Then some men live in
the Manor Cottages, and have their rents (ranging from
one and sixpence to three shillings) deducted from their
pay unless, like Croft and Lambert, they live rent free.
Quite a number, especially the three tractor men, put in
varying amounts of overtime. 'Piecework', such as hoeing,
'singling', and 'lifting' sugar beet, is seldom indulged in for
a whole week at a time, but has to be disentangled from
time worked 'by the hour'. And the weather is often responsible for the loss of a certain number of hours by those
who prefer dry clothes to a full week's wage.

Marsham generally brings in the outstanding time sheets
only on Friday mornings, in order to include Thursday
night's overtime, so the actual working out of the paybook
has to be left until payday morning. Though a guess at the

total must be made earlier, preferably the night before, and communicated to Major Rivers, who writes out wages cheque and cashes it with his own at the bank at Meddenham, bringing the Manor money round with him on his way back. And with the post and odds and ends disposed of, one is theoretically free to concentrate on the paybook itself.

Nowadays paydays have shed their more nightmarish qualities. The working-out process seldom occupies as much as an hour. And I have jettisoned the old system . . . which meant my sitting freezing in the outdoor office in the 'pumpyard' for anything up to two hours whilst the men came in to be paid, those working late invariably arriving the second when I attempted to reach the flat for a thaw or a longed-for cup of tea . . . and now pack the wages into the pay envelopes within range of the indoor office electric stove, and pay them over on my afternoon round. But during my first weeks at Silford I was far too busy trying to tread in the ordinary routine paths without slipping, to be able to think of short cuts. Nor had I managed to achieve the species of split personality necessary to deal with the eternal interruptions and emergencies which seem to be a feature of any farm manager's life.

"Please, miss, Barrett's just brought another five tons of pulp in. Here's the ticket."

"Is that Ashacre 12? Spitalfields wants you."

"Please, miss, will you ring up the Petroleum Board and order another three hundred gallons of paraffin."

"Did the Guv'nor leave any message for Will, where he is to go next?"

"The station-master says there won't be any more trucks till Monday."

"Are you Ashacre 12? A telegram from Manchester."

"Do you know what time the Guv'nor's coming round, as I want a word with him?"

"Barrett's never brought no sacks back last night, and if he hasn't left any at home, we shan't have enow Gardiners for today!"

"Please, miss, can I have some nails . . . ploughshares . . . nuts and bolts . . . cow's relief . . . a new broom . . . top bonds . . . rubber aprons . . . horse balls . . . cartridges for crow scaring . . . . or whatever it is, from the store or the veterinary cupboard?"

"Please, miss, could you have a look at my hand . . . arm . . . face . . . foot . . . eye . . . ?" For not only must you endeavour to be a cropping expert, a dietitian and concocter of balanced rations for stock, an engineer, a vet surgeon, and fifty other things besides, but a first aid and ambulance unit as well. And before I had been at the Manor a month, I had dealt with half-a-dozen cut or festered fingers, a foot which its owner had carelessly set on the point of a rusty nail, a hand which had had to be conveyed (with its attendant highly frightened body) into Meddenham to be stitched up by the doctor, an eye full of lime-wash, and the various cuts and scratches sustained by Patfield, who had bicycled into a fallen bough on his way to work one morning in the blackout.

They occur at any moment of each and every day, with a tendency to reach a peak on Fridays. So that in my early Silford days I counted myself skilful if the paybook was complete by the time I was due to establish myself in the outdoor office behind the long creosote-stained table (Peacock's handiwork, as he proudly told me), although so far no other payday has attempted to produce 'incidents' on such an epic scale as those provided by the past Friday in my first November.

The day opened innocently enough.

In fact, it began rather well, for when I rang up the Major with an early query from Marsham, his reply concluded with a complimentary reference to my progress and the news that he would be much too busy to get round to help me with the letters that morning, but that he was sure that I could manage them perfectly well by myself now, and so he proposed to leave me to it from now on. If there was anything I wanted to know, I could always give him a ring.

There was a touch of frost in the air as I discussed the carrot situation with Patfield in the yard before breakfast. Not enough to chill the blood, but sufficient to set it tingling and make me welcome the appetizing smell of grilling bacon which floated out to meet me as I neared the house, in spite of the long tradition of fruit and rusk breakfasts which lay behind me.

That bacon, though cooked for, was never to be eaten by me, alas! Before I had taken three steps towards the stairs, Mrs Dack came pattering down the back hall towards me.

"Please, m'm, there's Howard want to see you at the back door."

Reluctantly I turned and followed her.

Obadiah Howard (known to everyone on the farm, bar Mrs Dack, as 'Oby') was the pig-tender who looked after the inhabitants of the model piggeries at the top of the ten acre 'Pig-Pasture', and he combined a shortish plumpish figure and a broad cream-and-pink face strikingly reminiscent of one of his charges, with a maddeningly unhurried and unhurriable mode of speech and locomotion which nearly drove me to frenzy when I took down his weekly reports, or accompanied him from house to house on my visits of inspection. I should be lucky if I could prise his wants out of him in less than ten minutes.

"Them sows in Number Fourteen and Eight," he began, at the rate of perhaps a dozen words per minute.

"The two who've been off their feed these last two days, aren't they any better?" I interrupted, hoping to expedite matters.

"No, that they aren't. And there's three out of the eleven young'uns dead in Eight, and one in Fourteen, with another what looks like a goner. And the Wessex in Three and Five and Six haven't so much as been nigh their troughs yet."

As I already knew, it is a serious matter when an animal refuses to eat, to say nothing of the losses. But worse was to come.

"If you ask me what's the matter with 'em, it's swine fever," chimed in Marsham cheerfully, who had joined the party unobserved. "I've just been havin' a look at 'em. I can't say as I'm surprised. It's been a wonder to me as we haven't been let in afore, considerin' the amount there's been roundabout these last few months. Joneses, Martins, Burnetts, Jordans, they've all on 'em had it. I reckon you'll have to ring up the Guv'nor and ask if you hadn't better tell the police."

"Wait until I come round," directed the Major, when he heard the news. "I'll call on my way to the bank."

Mrs Dack had slipped the post in front of me whilst I was telephoning, but letters could wait until after breakfast. Removing the one addressed to me from the pile, I started for the kitchen to say that I was ready.

"Please, miss, you'd better come and have a look at Mark."

Marsham was at the office door before I could get out of it.

Outside in the drive Mark Buxton, the ace tractor-driver, was clutching his right wrist despairingly.

"Sent me flying clean over the fence, she did, when I was trying to start her up again just now, down by the washer."

"Reckon she thought you ought to be in the Air Force, and was givin' you a lesson in flying," contributed Marsham, with a broad grin. "If she ain't broke yer wrist for you, it's a rum'un to me."

"She" was the orange 'Fordson' which did all the road work, and was the pride of Mark's heart. And if she hadn't broken the bone, which seemed highly probable, she had certainly managed to sprain it pretty badly; and though I could apply temporary relief with hot and cold water bandages, it was obviously a case for the Meddenham doctor.

Leaving him with his hand under the scullery tap, I made for the back stairs to fetch the first-aid box from the flat, only to be stopped by Mrs Dack as I was halfway across the kitchen. But not the familiar aproned Mrs Dack wearing the usual half-plaintive, half-resigned expression which always preceded the query as to when I would be ready for my meal, but a set-faced stranger in a crushed shapeless black felt hat and a long black coat. In one hand she clasped

a battered brown suitcase. In the other was an opened letter
which she held out to me.

"It's me daughter Elsie, miss," she said, her under-lip
trembling. "I hate to leave you in the lurch. But I've got to
go at once, and if I start now I've just time to walk to Dennington to catch the Leeds train. I've left a note for the
baker on the windowsill. And you'll find the grocery cupboard keys on the mantelpiece."

In spite of a reticence that would have shamed an oyster,
Mrs Dack had contrived to let fall quite a lot about Elsie,
that paragon of only children, whose sole fault was that she
had chosen to wed "that Jim Bowthorpe, and go up North
to live, instead of taking Tom Jarvis, what earned as good
money, if not better, when she could have had a home in
Norwich near me."

Now Elsie had got herself knocked down by a lorry and
was in Leeds Infirmary in what the writer, who was Elsie's
sister-in-law, and evidently no graduate of the school of
thought which believes in breaking things gently, but all in
favour of the crash-bang knock-down method, had no hesi-
tation in describing as a very poor way indeed. She had offered to provide Jim with meals and give an eye to Merle
(aged eighteen months) till Mrs Dack could get there. But
what with her bad leg, and Jimmy away from school sicken-
ing for something, and her fifth on the way, she really
couldn't undertake to be responsible for more than a day or
two, and felt that Mrs Dack should come as soon as possible,
or there was no telling what might happen.

Mrs Dack was in no mood for sympathy, which was just
as well, since there was no time for more than a word or

two, and to say that there was no need for her to walk the
two miles to Dennington, as I could drop her at Banwell,
the station before, on my way into Meddenham with Mark.
Then I found the bandages, bound up the injured limb,
and started a cold and spitting Amy in the motor house.

At Banwell I deposited a humid-eyed but controlled Mrs
Dack at the bottom of the station steps, and then went on
to Meddenham, where the doctor examined Mark and finally pronounced a verdict of sprained only. But the wrist
must not be used for at least a fortnight, and come back
again on Monday.

"An' me been driving week in week out for the last ten
year, and never done nothing like it before," said Mark disconsolately, as I dropped him at the gate of the end red
brick box in the row of eight built by 'the council' at the top
of Ashacre Close. "Then you talk about it!" he added disgustedly, though in fact we had made the journey back
with scarcely a word.

As I was putting Amy back into the garage, the Major
drove into the yard, and together we marched off to the
piggeries.

"Haven't they even touched any of that Sussex Ground
Oats you got them from the mill yesterday?"

"Not a scrap."

"I don't like the look of it at all," he confessed, as we
went from sty to sty. "I'm afraid Marsham's right, though
I can't be sure. But you'd better ring up the Dennington
police and tell them we suspect it. Oby!"

"Yes, sir!" And Oby came up at twice his usual speed.

"Mix up a pail of disinfectant right away, and stand it

just inside the gate. Put in a fresh lot every morning, and you're to wash your boots in it every time you go in or out. You and me, too, Miss Harland. And no one else on the farm is to come on the Pig Pasture until further notice. If it *is* swine fever, it'll probably go through the lot, and you'll lose those weaners through the horseyard as well. But we might be able to save those thirty breeding sows up at the New Buildings. They're a good half-mile away. Tell Marsham to put someone else on to feeding them. Isbell, for instance. He lives down that way."

Accompanied by a strong smell of disinfectant and a now ravenous craving for food, I telephoned the police, who said that they would inform Lytham St. Annes and Norwich at once, and that the Inspector *might* be able to get out that afternoon. And at long last I reached the kitchen to see what I could do about breakfast.

The door of the electric cooker stood open, as it always did when Mrs Dack was keeping my food hot inside, lest it should burn, or the dishes crack. On the bottom shelf, still warm, I found a plate on which a smear of melted grease bore witness to the fact that a couple of rashers of bacon had once reposed upon its surface. Another patch of grease, congealed and cold, visible on the floor just by the oven door, mutely suggested its mode of exit. In the centre of the kitchen sofa, wrapt in the deep untroubled slumber known only to the conscienceless and the replete, was Master Samuel Pepys.

Plugging in the electric kettle, I made myself a cup of tea, and unearthed some bread, some butter and an apple.

"Butcher!"

On the back-door step stood a small boy with an enormous basket from which he produced the chops ordered by Mrs Dack for the weekend. And would there be anything for next week?

I suddenly realised that Mrs Dack's departure meant that I should probably have to grapple with the cooking and housekeeping as well. But this was no time for planning menus.

"I'll ring up on Monday," I compromised, and with a glance in the direction of the still sleeping Master Pepys, I locked the chops away in the larder.

The paybook!

But there is the post still untackled. My first day alone with it, too.

Six carrot returns to be booked up. (I book them.) Four cheques to be handed over to the Major when he gets back from Meddenham. An account of the last three loads of sugar beet sent in to the factory. (Top tare much too high. Must show this to Marsham and George Long, who is in charge of the sugar beet gang.) A trio of bills for the bill-basket. (Five shillings per ton discount on one if paid within twenty-eight days. Must see the Major does this.) A couple of circulars, one of which begs to draw my attention to the merits of Messrs Hoopers' sheep dip (there have been no sheep on the Manor for some years) while the other announces that on the following Thursday week a sale of pedigree cows will commence at ten o'clock, sharp, at a farm some two hundred and fifty miles away. (Just before he was called up, Captain Irstead bought a couple of heifers at a local sale presided over by the auctioneer whose name figures prominently on the front cover, and similar catalogues, featuring farms at an ever-increasing distance, have optimistically appeared bi-weekly ever since.) R. A. Leeds, of Spitalfields, are interested in parsnips and beetroot, if we have any to load (which we haven't). And the Manchester people are sorry to have to complain, but the large carrots are arriving badly scarred and the small ones none too clean. It looks as though they are being put through the washer too fast; and they will be much obliged if we will see to it.

This last means an immediate visit to the carrot-house, where Bushell says aggrievedly that if anything they're going through too slow. The real trouble is those big carrots off Pightles. The other fields are all right. But as long as we're washing Pightles, the only way to stop it is to run the washer to suit the medium large, and send down another man to pick the extra big ones out for the cows.

Marsham has to be found and consulted without delay.

"Don't want a man on that job," says the oracle scornfully. "One of the boys'll do for that. You didn't happen to notice if there was any crows on Booters as you come by, I daresay?"

I had, and there were.

"Mustn't take Cecil Peachey off scaring, then. Do them beggars'll have the lot. That's how they like wheat, with the green just showin'. But Harry Buxton's only jobbing about in the yard. If you'll tell him as you go back, he'll do."

Hardly have I dispatched Harry Buxton (Mark's elder boy, a lad of sixteen), and regained the office, than Major Rivers arrives with the wages money. Mercifully, however, he, too, is behindhand and in a hurry, and for once neither inclined for a farming walk or to suggest a cup of tea.

But before I get down to the paybook, it might be a good idea to write out the advice notes for the London carrots, which are going up by train today, and slip them into the letter box in the porch, which the 'Old Guv'nor' bullied the postal authorities into accepting as an official collecting box some twenty years ago. The blackout and other war conditions have altered the hour of the afternoon delivery and collection from four-forty to two o'clock, and it is now nearly half-past twelve.

"What's all this Charles has just been telling me about Mrs Dack running off to Leeds?" Mrs Rivers is on the telephone as I stamp the last envelope. "That Elsie of hers always was a perfect pest. I expect she'll be away for weeks, bother the woman. I don't know what to do about you. Mrs Heyhoe will probably come in in the mornings, instead of carrot picking. You'd better see her about it this afternoon. But I know she won't sleep in. You couldn't get hold of a friend who'd come and stay for a bit? If not, I expect you'd better sleep here, for I don't suppose you want to be in the house alone? Just as you were getting nicely settled in, too. I'd like to wring that Elsie's neck."

Most certainly none of my friends possessed either the time or the income to be able to take an immediate holiday of indefinite duration, though I registered the idea for future reference. It would be fun to put up M. or B. in the flat for a weekend or so if and when they could get away. For the obvious solution of spending my nights at Ashacre Hall, I felt no more enthusiastic than Mrs Rivers sounded.

Apart from the fact that you must be on the spot, I had grown very attached to the flat, and the privacy and independence that were mine whenever I had the time to linger in it.

As for being alone in the house. . . . Did the absence of Mrs Dack make any vast difference? In some ways it would actually be a relief, for whilst our relations were cordial on the whole, it was obvious that Mrs Dack privately regarded me as something of an interloper capable at any moment of committing assault and battery on 'the Master's' furniture. And the aura of patient suffering which hung about her almost visibly whenever I was called out or delayed just as meals were ready, though understandable, was almost as irritating as the interruptions which caused it. There was a heat plug beside my bed, for which I had already bought a kettle to make my morning tea. Croft usually lit the independent boiler and brought in the coal. And if Mrs Heyhoe, whom I already knew as a cheerful soul who fetched my washing on Mondays and returned it spotless and creaseless (for a ridiculously small sum) towards the end of the week, and who appeared to harbour none of the feelings of a faithful family retainer, though she was Marsham's sister and married to one of the men, would come in and do some cooking and tidying up each morning, I thought I could manage. Anyway, I could try.

"Well, you can always change your mind," said Mrs Rivers when I intimated as much. "And old mother Heyhoe's a jolly good cook. See you at lunch on Sunday if I don't call in before. G'bye."

I get out the paybook, write down the names, and have actually worked out several men's wages before there is a knock on the door. Lambert has chosen this auspicious moment to produce the cow record and feeding stuffs sheet, which he should have brought in the day before.

. As he departs, I catch sight of my watch, realise simultaneously that it is long past my lunch time and that I am again ravenously hungry. I cut more bread and butter in the kitchen, drink a glass of milk (which is soon to become my staple food, since it needs neither preparation nor planning out) and go back to the office leaving Master Pepys, who has watched the proceedings with one eye unclosed, to lapse into sleep again.

Kettering: overtime and forty-one hours in carrot washer, nine-and-threepence ha'penny. Lost one hour. Less rent two shillings. Baker: three hours overtime, less four days sugar beet. Mark: Sunday until ten at night repairing carrot washer, eight and a half hours overtime as well, a shilling toll paid at Sedgeham Market when delivering fat pigs. Bushell: Sunday repairing washer, carrot money, overtime. Isbell: two days drains, less rent. Meale: . . . only nine more to go. . . .

There is a ring at the front door bell.

Marsham, no doubt.

"Come in," I cry, without moving. Peacock: rent, two shillings, three days carpentering, lost Saturday morning. . . .

The bell rings again. It can't be Marsham, bother it. Banging the office door behind me, I cross the hall to the front door and find the Rectory on the step.

Mrs Rivers has already begun to launch me on the local social sea, so Mrs Cutbush and I have met before. Fair, fat, fiftyish, and a 'foreigner' like myself, the wife of the rector of Silford-with-Ashacre suffers an additional handicap in having Mrs Rivers at the Hall. Everyone in both parishes rallies to 'Miss Kate's' energetic standard, unanimously elects her President of the Women's Institute each year, knits feverishly for the W.V.S. Comforts Committee, has voted for the First Aid Point being established in the late power-plant house at the Hall instead of in the empty Rectory gardener's cottage, and invariably flies to the former for physical and spiritual consolation when difficulties arise. The only lay activity on which Mrs Cutbush has managed to pounce first (Mrs Rivers having forgotten the date and attended a Point-to-Point meeting instead) is the National Savings Group inaugurated last April. And since the war began, in her efforts to see that Ashacre and Silford lend more to defend their country than any other pair of parishes in England (and so wipe metaphorically for once at least her rival's eye) she harries the villagers day in and day out, when asked to contribute a whist drive prize does so with so many stamps, sits at the receipt of custom in the Rectory study from six P.M. onwards every Monday evening, and this afternoon has called to enlist my support in launching a savings campaign among the men on the Manor farm.

The flat is still fireless and undusted, my bed unmade. I usher her perforce into the office, not without hope that the sight of the littered desk will speak for itself. But where National Savings are concerned, Mrs Cutbush has become a fanatic, and is blind and deaf to all else.

Agricultural labourers' wages have just Gone Up. (To hear Mrs Cutbush, one would imagine that the rise is two pounds a week instead of two shillings.) There is a War On. This is a time of National Emergency, and Every Farthing Counts. So easy to deduct it before paying out their wages.

Fortunately the door-bell rings again before I am hypnotised into committing myself. This time it is Marsham, who says that the Inspector has received the police call in time after all, and here he is to look at them pigs.

I tell Mrs Cutbush that I am afraid I shall have to go, but that I'll speak to the men, and shoo her out of the office, lock the door (since the wages money, over a hundred pounds, is in an attaché case on the desk) and go out into the yard to find the Inspector.

A sandy haired direct Scot, he is as anxious to get his inspection over as I am to see him off. But inevitably it takes a certain amount of time. There is much washing of rubber boots in the disinfectant pail. The affected sows' temperatures are taken, including three new suspects of whom Oby reports that 'they don't look muchers'. This morning's corpses are disembowelled in order to subject portions to a laboratory test (during which proceedings I hastily revisit the occupant of the farthest sty). And having added a few supplementary instructions to those issued earlier by the Major, the Inspector finally leaves, saying that we shall hear from him in due course . . . confirming the disease, he fears.

Green: thirty-five hours in the carrot washer. Lambert: one calf. Croft as usual. Nine acres of sugar beet at thirty-eight and sixpence. And that finishes the paybook. Snatching up the attaché case, and suppressing thoughts of a cup of tea, I reach the door of the outside office a short head behind Bushell, the first arrival.

One by one the men straggle in.

Bowes has put in an extra hour carting straw and forgotten to tell Marsham about it. And what do I think about these here Finns? I have made a mistake in adding up somebody else's money, and hurriedly work it out afresh. Earlier in the week I have dressed a cut finger for Catton, one of the sugar beet lifters, and now he removes the leather 'hutkin' which protects it to show how well it has healed. Heyhoe kindly undertakes to see his wife about coming in tomorrow, and promises that she'll be along by nine sharp if I don't hear different in the meantime. The minutes slip by, and as the temperature of the outside office falls lower and lower, my thoughts stray ever more often in the direction of the kitchen and a cup of tea.

Only three more to pay, and I can see them coming into the pumpyard through the glass upper half of the office door, Lambert leading the way. My cup of tea. And it must be close to blackout time.

I turn to the paybook and have almost finished counting out Lambert's money, when there is a sound of running footsteps, and a perfect stranger bursts open the door.

"You want to ring up the fire brigade at once. There's a stack of yours on fire up the Ashacre road."

Scattering piles of copper and silver all over the table and the floor, I fly out of the office, only stopping to lock the door, and indoors to the telephone. "Which is the nearest fire brigade?" I demand of Marsham, who is on my heels.

"Meddenham. They're only four miles. And they come to that haystack at Yew Tree the other week."

Yew Tree Farm is in Silford, and less than a mile and a half away. I pounce on the telephone directory, then dial o.1. and demand Meddenham 215.

It seems an hour before they reply, and then it is only to say that they can't come. We belong to Sedgeham.

"But Sedgeham is eight miles away and you're only four. And you came to the one in Silford last month," I protest.

"And a nice lot of trouble we got ourselves into because of it," an aggrieved voice replies. So I replace the receiver, dial Norwich once more, find that I have not yet been rung off, start all over again, and finally get through to the Sedgeham fire brigade, which promises to come at once. "Bring an extra long hose," I add, as Marsham hisses in my ear. "The stack's at least half-a-mile from the river."

Ashacre Hall must be informed, I suppose. So I dial 57, when the maid answers, and tells me that the Master and Mistress are out, but shall be told directly they get back.

Everyone has melted away by the time I replace the receiver, except the stranger who, by one of those unlikely coincidences forbidden to novelists but comparatively common in real life, turns out to be an inspector of the insurance company which handles all the Manor farm insurances. And to stretch coincidence to its utmost limit, he was talking to the chief of the Sedgeham fire brigade barely half-an-hour ago. He is now on his way back to Norwich, and offers to drop me at the stack as he goes . . . an offer which, knowing Amy to be in one of her sulky fits, I gratefully accept.

He has already told the story of how he discovered the fire several times, but caught up in my spate of telephoning, I have heard only snatches of it. I am now able to hear it in full. By the look of it, it had only just caught when he pulled up, he avers. And two small female children, found weeping by the gate and promptly interrogated, had spoken of 'boys' and pointed up what I identified as the 'Sandy Loke' leading to Silford Hill.

The stack, situated at the bottom corner of 'Doles' bounded by the Ashacre road and Sandy Loke, is ablaze from end to end as we reach the scene, and in the road and along the lane fence has already gathered a crowd which is augmented every minute. Pedestrians of all ages and sex push forward from the rear, determined to miss nothing. Cyclists either fling their machines against the bank of 'Barn Breck' opposite, and join the crowd, or pedal furiously along the rutted inclines of the Sandy Loke, up which several of the men have already vanished to follow on the track of the one and only clue, the 'boys' reported by the two small children interviewed by Mr Balt, the insurance inspector. Cars draw up and have to be requested to park lower down the road (so as not to impede the progress of the fire engine when it comes), among them one from which springs a tall alert man of middle age, hatless and wearing a suit of tweed plus fours, but with 'police' written over every inch of him. (Sure enough, he introduces himself as the Westham sergeant.)

"Nice thing to happen when the blackout's on," he says: a sentiment promptly echoed by several bystanders, who hope audibly that 'they' haven't arranged to bomb Silford tonight, do they'd knock a hole or two in it (though nobody makes any attempt to leave on this account).

Speculations as to the fire's origin fly about as freely as the sparks which shower continually on the field and hedges.

Some of the schoolchildren must ha' been playing at Guy

Fawkes, and Lizzie and Alice Parden been as bad as any of
'em, for all they had to run crying home when that chap
came along in that car. It's a German spy, disguised as a
tramp: "I see him meself, yesterday, looking at a lot o'
papers he took outer his pockets." No, 'tain't. It's that poor
saunsey, Henny, what should have been put away years ago.
It is some evacuees, least that's what they looked like, and
they've been selling George Drayton's Radio Times and
newspapers, what they took off his bicycle when he left it
outside his gate earlier on, at half-price round the village,
when they'd taken a watch what a lorry driver'd left hanging on his steering wheel, when he stopped outside Parfitts',
as well.

Surprisingly enough, this last turns out to be the truth.
For a few seconds later Marsham is seen leading a procession of nine small boys followed by half-a-dozen men
wheeling bicycles down the lane. And on being cross-
examined by the Westham sergeant, who promptly takes
them in charge, they own to being truant inmates of an
evacuated charity home now at Sunnington House, some
eight miles distant. Taxed by an incensed George Drayton,
they also admit the removal of his papers, though they deny
knowledge of their present whereabouts or having received
money for them. They also deny taking the watch and the
apples.

Apples? What apples?

"Mine," says Joe Gidney, one of the dismounted rear-
guard, whose home is on Silford Hill. "I seed you take 'em.
And, what's more, I seed you hull 'em away into the ditch
the minute I start to come arter you."

Jumping on to his bicycle, he pedals rapidly away into
the dusk, to return triumphantly bearing a dinner bag full
of apples, the lorry driver's watch, and three boxes of
matches, just as the culprits finish their version of the firing . . . an ingenious if somewhat unveracious story which
ignores sundry footmarks found in the soft earth around
the stack-bottom, and relates the extraordinary behaviour
of two solitary matches found lying loose upon the road
which had been struck (on the breeze, presumably) and
then thrown over the fence, when they had sailed some
twenty-five yards through the air and landed, still alight, on
the stack.

With the advent of this fresh evidence, the cross-exami-
nation is re-opened, when eight of the culprits unite in
blaming the ninth; and the Westham sergeant, whose car
is an 'occasional four', decides that it will be simpler to accept this part of the story for the present, parks the criminal
in the back of his Morris-Eight, and asks if he may use the
Manor telephone to ring up the Meddenham inspector.

We adjourn to the Manor office, when the Meddenham
inspector is informed of the crime, and says he will be over
right away, and I ring up Sunnington House to ask have
they lost any evacuees as we have nine here? Yes, they have,
and will send a van to collect them as soon as it gets back,
which will be any time now.

Back to the stack, to find the crowd has been enlarged by
the addition of Major and Mrs Rivers on their way back
from Norwich, and the fire brigade. Both the former have
quite an amount to say, and are busy saying it, but I have no
time to listen to more than a few words before the fire bri-

gade announce that, in spite of my warning, they haven't brought enough hose. May they borrow the Manor telephone, too?

They do, after depositing the fire engine in the yard (they will never be allowed enough to reach the river — the best they can do is to use the Manor pit and hope it won't run dry too soon). But it is obvious that no hose can come along for something like half-an-hour, and it is now seven o'clock, and my craving for a cup of tea eclipses every other consideration. Would they care to join me? I enquire, as I plug in the kitchen kettle.

They don't mind if they do, so for the next twenty minutes we sit in the kitchen drinking tea from Jubilee mugs (somewhere, no doubt, there are plenty more cups and saucers, but their whereabouts at present are known only to the absent Mrs Dack, and the mugs left over from the Silford Jubilee frolic are the only things I can find), eating slices of Mrs Dack's excellent fruit cake, and discussing the war, fires, and evacuees, of which latter the third fireman contributes a racy description of the scene at last week's Sedgeham market (when a free fight took place with eggs as ammunition), and an account of the domestic habits of the three taken in by his married sister, but we agree that they're all sorts, and of course you always hear more about the bad 'uns.

The telephone rings, and Sunnington House enquires where exactly are we, as the van driver is now about to start, but doesn't know the way. Neither do I, but fortunately the second fireman does. After which we finish our tea and most of the cake, and express sympathy with Number One fireman, who had, he says wistfully, planned rather a different evening for himself. (He's not the only one.)

The hose arrives, the firemen troop back into the yard, and I take a torch and grope for the abandoned wages money and paybook in the outdoor office, scurrying back to answer the telephone, which is Major Rivers asking what they're up to now? And have the evacuees been collected yet?

No, they haven't. For as I cross the road to the Sandy Loke, I see them, a forlorn pathetic row silhouetted against the firelight. Shall I be compounding a felony if I walk back to the Manor and return with the remainder of Mrs Dack's fruit cake? But I shall never know, for at this moment the van arrives, and all nine . . . for the Meddenham inspector says he's got nowhere to lock up the arch-criminal for the night . . . are bundled into it and are whisked away.

At last I gain the shelter of the 'flat'. As I am about to tumble into my hastily made bed, I remember my letter, still unread, and rescue it from the pocket of my jodhpurs. London is hectic just now, writes M., what with the refugees, the sirens, and the blackout. And now A. (her husband) is talking about volunteering to help the Finns. How she envies me the peace and quiet of the country, where nothing ever happens.

Sometime in the small hours I woke, and looked at my watch. Three o'clock.

From the direction of the pit I could still hear the clanking of the fire-engine, and the occasional protest of an irate duck.

≍ IV ≍

AS the days continued to race by at a speed they had never in my life approached before, the work began to fall into line a little, the separate fields to take shape. The Manor, built to replace the original house which had been burnt some eighty years before, presented a most unlovely phase of Victorian domestic architecture, but even that now acquired for me a not unpleasing homeliness, and the people around it became persons.

There had been forty-three names in the Manor labour book just before the war began. But two of the five sons of Long, the second cowman, Herbert Bushell, brother of the carrot washer foreman, and three of the latest generation of the Buxton family (whose members have worked, man and boy, on the Manor farm for the last hundred and fifty years or more, and who are related by blood, marriage or both to practically every family in Silford and Ashacre) had been seduced by their master's eloquence at post-Crisis recruiting meetings into joining the Territorials; and so they had left for their regiment when he did . . . though so far only Arthur Long, now on his way to India, his father proudly told me, had left the country. The others had been drafted into a newly formed battalion temporarily billeted in Sedgeham, where they had been joined by Peachey's eldest boy and another Buxton just before my arrival, and now were often seen hanging singly about the yard or in pairs, exchanging pleasantries with the Silford belles along the Ashacre road.

Another pair, Jeremy Peacock (the 'estate carpenter's' nephew) and Leslie Long, the third tractor driver, went before the month was out. And Sidney Peachey and 'Oby' were due to register in January. But Major Rivers thought that now the powers-that-be were talking of lowering the reserved age for agriculture, it was unlikely that either of these last would be called up. And by an intensive bombardment, at long and short range, of any authority he could reach, he eventually secured the release of Leslie Long early in December. Whilst Tom Buxton, father of Mark and Vincent, who had retired four years back on the Old Age Pension and a modest win in a football pool, had come forward to offer his services in any and every capacity. So that while Marsham grumbled at being short handed, and sundry odd jobs were neglected, the labour shortage at Manor farm was not yet acute.

In the beginning I had wondered a little about the men . . . how they would react to taking orders from a woman, a 'townee', and a complete ignoramus into the bargain. I had also heard a good deal about the notorious Norfolk attitude to 'foreigners', a compound (so I was told) of distrust and suspicion warranted not to dissolve under twenty years, if then. But I need not have worried about either. Human nature is very much the same wherever you go. And you seldom meet with a really unfriendly reception unless you go out of your way to look for it. As for the legendary Norfolk reserve, I have yet to meet it. Isbell's occasional attacks of moodiness and Lambert's disgruntled fits are no more indigenous to the soil than the swallow is a native of any one country. And as far as the rest are concerned. . . . Talk about women being the gossiping sex. They simply aren't in it compared with men.

The Manor men are willing at any time to discuss the latest B.B.C. news, the responsibility for the present state of world affairs, the crops, the weather, and themselves. If I allow myself to be caught into conversation within sight of leaving-off time, I can seldom escape in less than a quarter of an hour, and then only by brutally scissoring the thread of a half-told tale by resolutely turning my back on the teller. Whilst during working hours I can only get away by sternly reminding them (and myself) that quantities of work are waiting for me elsewhere, and that it would never do for me to invite a charge of wasting their time from Major Rivers or Marsham . . . though, as a matter of fact, both the latter are every bit as bad (or as good) when it comes to telling a story.

There was plenty to talk about in the chill December days which followed the fire.

The King returned from a four day visit to France, having been escorted over a sector of the Maginot Line whose impassability was still, at that time, one of the first Allied Articles of Faith ("Just you wait and see what happens when Jerry starts to attack. They'll never get through T H A T," one or other of the men was bound to contribute to any discussion of the war news). The offending evacuee

was brought to Meddenham in a police car and dealt with
by a specially summoned Juvenile Court, Major Rivers re-
tiring on this occasion and leaving the case to his fellow
magistrates: while Mrs Heyhoe (now installed at the
Manor from nine till twelve six mornings a week) delivered
the lay verdict when she expressed the pious hope that
somebody had warmed the jackets of the remaining eight
"do we shall all on us be burnt in our beds one of these fine
days. Though what can you expect, browt up in one a' them
homes? An' one on 'em saw his father cut his mother's
throat afore they took him away, so they tell me." The
Soviet Government refused the League of Nations offer of
mediation in the Russo-Finnish dispute, and the Finns continued to fight stubbornly on, with surprising success. Sidegate Breck was cleared of carrots, with the gratifying average of twenty-three tons to the acre, and Vincent Buxton
was able to riffle it up right away. A letter from Leeds announced that Elsie was going on as well as could be expected, but her mother couldn't rightly say how long it
would be before she was about again: meanwhile I would
find another cake in the square tin on the top shelf over the
dresser, the scrubbing soap was kept at the back of the stick-
cupboard, the Sunday paper had been paid for up till last
week, and the writer was mine respectively, B. Dack. His
Majesty's ships Ajax, Achilles, and Exeter chased the Graf
Spee into Montevideo harbour, standing by until she came
out and sank herself four days later, and Mark, nominally
off the premises, but actually spending most of his time
prowling disconsolately around with one arm in a sling and
both eyes and ears wide open for any sign or sound of Vincent, Peel, or Bushell (who is sometimes pressed into service

s a driver) maltreating his tractor, cheered up for the first
ime since his accident, and presented me, beaming all over
is seamed and weather scarred face, with a cartoon from
is newspaper depicting Hitler pinning on the breast of yet
nother of his captains, the Order of the Scuttle, in the
hape of a miniature coal bucket. And swine fever, its exist-
nce confirmed both by the Scots inspector and the death
f one pig after another, continued to run its desolating
ourse.

The Wendeswell policeman, who had nailed up the offi-
ial Order concerning Affected Premises on the garage
oor, and whose duty it was to attend each and every fu-
eral to ensure that the corpses were well and truly interred
with a sufficiency of lime, grew tired of his enforced daily
isits, and begged me to institute mass burials twice a week.
The small 'suckers', free from the disease itself but faced
with starvation as their mothers sickened, were disposed of,
with the surviving 'weaners' in the horseyard, to the local
utcher. Finally the Major, remarking that it was no good
going between the bark and the tree, and with feeding
tuffs so difficult to get (and he had been told that the posi-
ion would get worse) pigs were rapidly becoming a liabil-
ty instead of an asset, obtained the necessary licence and
old every pig on the farm. Whilst Oby, offered the choice
f taking over the care of the Silford Hill cattle, or reverting
o the status of ordinary farm labourer, decided that he
would prefer his Sundays off for a change, and was set to
work cleaning out, swilling down, cementing up rat holes,
whitewashing, creosoting and generally disinfecting the
piggeries.

The shooting parties which had been a feature of autumn at the Manor and Ashacre Hall . . . the farms do not actually touch, but the Major and his brother-in-law hire the sporting rights over the intervening land, and in pre-war times organized four or five joint 'days' when almost the entire personnel of both places turned out to act as 'beaters' . . . had been abandoned this year, since, as the Major sadly remarked when recounting the glories of past 'days', Nazis had 'come in' on September the first instead of partridges. And with most of the usual guns limited to short, often-cancelled and never-coinciding periods of leave (if they could get any leave at all) it was no use attempting to arrange a shoot. But he sometimes spent an afternoon 'walking up' accompanied by Myhill, the keeper, his own retriever (an immensely dignified golden Labrador called Buster) and Nicholas Irstead's black spaniel, which occasionally answered to the name of 'Rags' but more often than not capered excitedly before and behind the party until put on his lead by Myhill. And on these occasions he seldom failed to leave some of the spoils, a pheasant, or a brace of young 'Frenchmen', in the Manor game larder.

"You can tell the young birds from the old ones by holding them up by the lower half of their beaks," he explained one afternoon, pushing one of the limp, soft feathered bodies into my hands. "If it bends when it takes the weight, it's a youngster, like that one you've got there. Look at this, though." And he held up a second. "Doesn't give at all. He wasn't hatched this year, or last, either, I shouldn't wonder. A brace be enough for you? I shan't be shooting again until the end of next week. But you can always tell Myhill if you can do with a pigeon or a bunny."

"But why 'Frenchmen'?" I asked Croft a morning or two later when, tracking him down to deliver a message from Mrs Rivers about lavender bushes, I discovered him sitting on an upturned box in the 'copper-hole', plucking one of the sandy-coloured birds. "Do they migrate here from France? Did the breed originate over there? Or what?"

Croft's ginger moustache twitched delightedly.

"See the colour a' them legs?"

He held them up, a deep rich crimson.

"Same colour as the French soldiers used to have their trousers made on, or so they say," said he. "And they reckon as how they used to be able to retreat quicker'n any others. The partridges aren't much for flying. But can they run! Not that we can't do a bit in that line ourselves when we give our mind to it," he added, with an air of strict impartiality. "I remember in the last war I thanked my lucky stars more'n once as I'd won the quarter mile three years running at the Ashacre Whitsun sports. That was in the spring of '16, when things weren't going any too well where we was. First we got the order to retreat and then it was 'every man for himself'. Me and about a dozen more was moving along, and not so slow at that, when bang! and afore I know where I am, I'm lying on my back at the bottom of a shell hole.

"I was a bit dazed like for half-a-mo. Then out I come and have a look round. There's all me pals in a row, laid out like a lot of ninepins and 'sdead as doornails.

" 'Here's a nice go,' I thinks to meself. Still, it wasn't no good me thinking about burying on 'em, with Jerry just behind, and shells bursting all round. So I crawls foward into a trench, empty of course, and kept agoin' till I come to the officers' quarters. No one there, neither, but they'd left quite a bit of gear behind. Pity to waste all that rum, I thinks to myself. So I took a look round, and found up three or four water bottles, and filled 'em up, ate some biscuits what had got left behind in a tin, and off I goes on me travels agin. But me luck was out this time, though in some ways I suppose you could say it was still in, in a manner o' speaking. Round the corner of a communication trench, just as I was doin' nicely, come a party of 'em, and one had just upped his hand to chuck a hand grenade at me, when the officer stopped him.

"I can see him now," Croft went on dreamily, his finger still mechanically stripping the partridge. "More like a bloomin' bally-dancer than an officer. 'Come, Tommy, come,' says he, mincing along towards me on his toes. So I come. Couldn't very well do much else, could I? Still, he let me keep the rum. I reckon he thought it was water. I wasn't half popular with the chaps when they slung me in amongst the rest of 'em.

"After that it wasn't so good. There were four hundred of us and a horse in a wire-netting pen nothing near the size of the Walnut Meadow, and all we got to eat was one red herring between four of us and a lot of uncooked fowls' grub. Though we each had a cup and saucer and plate to eat off after the first month, made us by a Belgian woman what had a pottery close by, and offered to do it, and as she'd offered to do the Jerries some and all, they let her.

"The horse was supposed to live on the grass in the pen.

He did worse'n any of us, and we weren't exactly gorging ourselves. Every day there seemed to be a fresh rib sticking out, until one morning he gave up trying. I had a bit of one hoof, but I can't say I got much benefit out of it. There wasn't a square meal for one on the whole of him. Next day the Jerries began to retreat, and it turned out that the officer in charge of our lot had been counting on Percy, as we used to call him, to pull the wagon with his loot on it across Belgium. He didn't half mob and go on when he heard what had happened, though poor old Perce couldn't ha' pulled a pram a couple of yards, let alone all old Beeryface's goods and chattels what he'd been pinching out of the chateaux.

"Still, there it was. And as we'd eaten his horse, he said, now we could pull the wagon instead. So he harnessed sixteen of us into it, and off we had to go. Five days of it, we had, and then they signed the Armistice, so they turned us loose to make our way back across Belgium as best we could.

"Proper little heroes we was in them days. Every village you come to, there was all the people cheering and going on, and offering you a litre or more of soup and half a loaf of bread. I had to give the soup a miss sometimes after the first five I stopped at. Some of those villages were pretty close together, and I was always a good walker. But I never said no to the bread, so I was carrying quite a packet by the time I got into our lines. And I've still got the cup and plate and saucer. I'll bring 'em in for you to look at, if you like."

"Haf you a minute, miss?"

I turned to find Marsham in the doorway.

"Were you doing anything pertickler this morning?"

"Not for half an hour or so. What is it you want?"

"Well, you see, it's like this here."

We were to start chopping out sugar beet on the lower half of Three Bridges, the long field adjoining the Three Bridge meadow tomorrow. And when it came to be carted off, the best place to stack it, both from our point of view and that of the lorry-drivers who had to load it and cart it away, would be in a disused gateway in the fence on the other side of the road . . . which didn't belong to us.

Normally permission to make use of their fences or gateways for this purpose is granted by neighbouring farmers as a matter of course, and is reciprocal. But in this case there was a stumbling block.

Old Jimmy Reeve was a rum client, said Marsham thoughtfully. He and the Old Guv'nor had fallen out over something, and Mr Nicholas had finished the job by refusing to let the engine stop to thresh Jimmy's barley the last time Three Bridges was oats, and the Manor set threshing it out. While that yellow dog of the Guv'nor's had set about Jimmy's hens the Michaelmas afore last. Leastways that was what Jimmy said, though the Guv'nor wouldn't have it. But Jimmy always bought a poppy off Miss Kate each Armistice. So perhaps if I was to go round and ask him . . . I couldn't mistake it. The house was down that track, you couldn't call it anything else, exactly opposite the boundary fence. And it would be a proper godsend, what with the carrots blocking up both the gateways this end, and no lorry couldn't get along the meadow road.

I'm sure the Brothers Grimm made a mistake when they said there were only seven dwarfs. Once upon a time there must have been eight. But whereas the seven in the story

stayed on their mountain to find first Snow White and then
fame in a Disney cartoon, the eighth was transported to
Silford, where he married and settled down to farm the little holding down at Silford Hole about the time the Wicked
Queen was doctoring the apple, and there he has been ever
since. There is no other possible explanation for Jimmy
Reeve.

Leaving Amy by the roadside I followed the grassy rutted
track described by Marsham for about a hundred yards,
when it turned sharp right to avoid a beech wood, curved
left round a tiny overgrown garden, and ended in a small
muddy square, flanked on two sides by a collection of toy
wood and thatch buildings in the last stages of dilapidation,
open on a third to the climbing slopes of a wired-in rabbit
warren, whilst on my right, scarcely bigger than the large
barrel lying sideways and evidently the home of an enormous black and tan mongrel, now barking furiously at the
end of his chain, was the back of a tiny house whose front,
if front there was, was completely hidden by the thicket of
brambles and bushes which had half filled the garden I had
just passed.

Carefully keeping out of reach of the now half-frantic
dog, I edged my way through the mud and knocked on the
open back door. Inside I could hear scufflings and bustlings,
but it was not until I had knocked a second and third time
that a little person, Mrs Tiggy-Winkle in every particular
except the prickles, bobbed out from behind the door.

Oh, I was looking for Mr Reeve, was I?

Yes, please. Was he anywhere about?

Mrs Tiggy-Winkle couldn't rightly say, but he should be

in the cowshed milking the cows. Straight through the garden and I couldn't miss it.

The garden was backed by a tall thorn fence, beyond which lay a small meadow, whilst immediately to my left was a low brick and tile building about twelve feet long. There was no sign of a doorway in it, however. Nor did my tentative cries wake so much as an echo. So I decided to explore farther.

At the far end of the meadow, near two cornstacks of Noah's ark dimensions, I came on what for want of an appropriate name I suppose I must call a cattle yard. Its proportions were certainly reminiscent of one half of the buildings at Silford Hill. But its appearance outdistanced anything that Mr Disney ever thought out. A cross between a gigantic mushroom and a native kraal, a bird's nest and a Robinson Crusoe nightmare, its roof boasted a thatch so weatherstained and patched that only magic could account for its contriving to stay on or keep out the rain. Three of its sides were a mixture of gorse, ferns, branches, and decaying sacks. The fourth consisted only of a few loose rails, behind which could be seen a floor raised by the non-removal of countless years' litter and manure, to some twelve inches above the meadow level. In one corner a pile of mangolds (large beets used for cattle feed) was dumped beside a small heap of straw on which a couple of hens were scratching dispiritedly. Behind them, on the shafts of an ancient tumbril, roosted two or three more. Leaning over the topmost rail, an elderly white horse regarded me without a gleam of surprise or interest in his lack-lustre eyes. Of cows, of other sheds, or of Mr James Reeve, there was no sign whatever.

"Mr Reeve. Mr Reeve! Mr Reeve!!!"

As I was about to give it up and turn back towards the garden, something stirred in the blackness beyond the tumbril, and a moment later Dopey's lost brother emerged, blinking, a milk pail in one gnarled hand.

"Mr Reeve?"

"Ar. And what met you want, miss?"

I explained my errand.

"Ar," remarked Mr Reeve meditatively, and settled down to chew the idea over in his mind with the deliberation of a cow with the cud.

"Mr Irstead, he's away in France, they tell me," he ejected at length.

"Yes." I had very nearly replied "Ar."

"Well, I don't know as I've got aught agin it."

"It's awfully kind of you."

"You might tell the Major as he's never paid me for them hens what that dog of his'n killed two years ago," rejoined Dopey II, suddenly alert, and dived back into his cavern before I could say another word.

Before, as Marsham phrased it, there was time to turn round, Christmas was on us.

Since Christmas Day fell on a Monday and all the carrot salesmen had written or telegraphed not to load over the holidays, the Major suggested that I should spend the long weekend in London, which I did.

It was good to be back, I told myself that first evening, as I joined in the animated discussions that are a feature of B's parties. But this was better still, I found myself thinking as I stood by the open window of the flat the following

Wednesday evening, breathing in the fresh cold air and watching the searchlights playing noughts-and-crosses in the sky.

There had been one or two sharp frosts over Christmas, alternating with sudden thaws. During the next few days, however, the frost held most of the time, and by Monday the ground was iron hard. No sugar beet could be touched although there were five acres still to lift, and the factory was due to close in less than a fortnight. London and Manchester were optimistically telegraphing, 'Load heavily', but of course carrot picking was equally out of the question, and we had already washed and despatched those stored in the house. And ploughing, whether by horse or tractors, was at a stand-still.

"I'm afraid this is going to last," the Major announced when he called round again soon after lunch. "So you might as well thresh out a stack or two. There's nothing much else you can get on with whilst this frost holds. And it's time you got some more corn out. Where are Peel and Catton?"

"Trimming fences at the bottom of Coney Hill."

"Then you'd better stop them. Peel must get up steam on the engine. And they've to set her against that barley stack on Crabb's Castle, ready to thresh first thing tomorrow, tell Marsham."

The barley stack on Crabb's Castle was an old friend, for one of my duties is to inspect every stack on the farm immediately after a gale, in case any thatch has blown loose. And this particular stack, though standing in what appeared to be a comparatively sheltered spot at the bottom of Silford Hill, had twice been a casualty. It had also led to a temporary coolness between Bushell, who had thatched it,

and Marsham, the latter scathingly and somewhat loudly
opining that Walter wasn't half the thatcher his brother
Herbert wor. Whilst Bushell himself came down from his
ladder with an air of a much injured and misunderstood
man, to confide that he didn't reckon nothing of them
thatching needles what Mr Nicholas bowt last year . . .
there'd never been none of this trouble time he'd pegged it
down with brotches.

As the Major stopped his car by the Silford Hill buildings
next morning, the drone of the engine floated up the hill
towards us.

"Hear the steady sound she's making?" he asked, as we
neared the field gate. "That's the way she should carry on
all the time. If she suddenly begins to chuck, you know
they've let a sheaf in without cutting the string first. Or
there's something wrong somewhere."

Round about the threshers a miniature dust storm was
raging. Already the white frost crystals which carpeted the
ground elsewhere had been overlaid with a thick greyish-
brown coating of it. So had Isbell's dark face as he stood
stacking up the corn. And it was several minutes before I
realised that the goggled, be-littered figure sacking chaff
and 'coulder' on the far side of the drum was Oby.

Up on the stack Ted Long, Meale and Heyhoe were
pitching the corn into the drum which gulps it down and
somewhere in the capacious depths of its rattling interior
separates chaff, grain and straw, ejecting the two former
into the appropriate sacks awaiting them, whilst the straw
is thrown onto the straw pitcher (or elevator) and forms
another stack.

. The Major came to a halt beside this latter, and stood for a moment watching Kirk and Wire, who were stowing away the straw as it fell, into a rough rectangle.

"Two's enough on this job as it's barley," he commented. "But you'll want three tomorrow as it'll be wheat, remember. So there'll be twice as much straw. Looks an easy enough job, doesn't it?"

It did.

"So it is," the Major went on with a reminiscent smile. "The only thing you mustn't forget is to put some under your feet occasionally, or else keep moving about the stack, if ever you have a shot at it. Geoffrey Blake thought he'd show the chaps how it should be done last spring. I happened to come round with Nicholas when the stack was three parts up. All the men were practically helpless with laughter, though those on the corn stack were still putting sheaves through as fast as they could. The pair on the straw stack were leaning on their forks watching what looked like a fountain of straw playing just under the elevator . . . which was Geoffrey pitching away for dear life to save himself from being buried. He was nearly eight feet down when we stopped the engine."

Dodging under the whirling belt connecting engine and drum, we stopped beside Isbell to take a sample of the barley as it ran into the sack.

"Still on the rakings by the look of it," the Major said, as he pointed out several small stones among the corn kernels. "Bit damp, too." He emptied his hand into one of the small paper cash bags we had brought with us, and I carefully pencilled 'Rakings' on it before putting it into my coat pocket. "I must be getting on. But they ought to be on to the main stuff any second now. So you had better hang on for a few minutes for another sample. And I shall want about three more taken at intervals during the day. I can have a look at them tomorrow morning. You won't see me again today. I've got to go into Sedgeham before lunch. And I've got a Drainage Board meeting on this afternoon. But you can give me a ring after tea to let me know how many coombs you've got out. There ought to be about a hundred and twenty."

A coomb is a sackful, and the weight varies according to the kind of grain, wheat being eighteen stone, barley sixteen, and oats ten. On my next visit to the stack I was invited to improve on the Major's estimate by Marsham, who reckoned there was nigh on a hundred and thirty. Counting the number of full sacks already out and taking another look at the remaining portion of stack, I hopefully hazarded a hundred and twenty-one, and surprisingly landed within two coombs of the actual total. Attempting to repeat this triumph with the Doves Close wheat, threshed the following day, however, I was no less than thirty coombs out. And my guess at the total of the Lincoln's Field barley, threshed the day after, was little better. Nor could I pride myself on my performance when it came to comparing the respective merits of the two barleys.

"Always look at barley in a north light," the Major told me, leading the way from the office to the front doorstep. And taking some from each bag in either hand, he held it towards me. "There you are. That Crabb's Castle barley isn't a patch on the Lincoln's. Look at it."

I looked.

"The Crabb's Castle's inclined to be steely, see? And look at these unripe kernels, too. I told Nicholas he'd cut it too soon. It ought to have had another four or five days. I'll bet there's a difference of at least ten shillings a quarter in it."

Whilst the Major elaborated his theme, I went on looking.

Nowadays I am beginning to notice the infinite variety in any grain. I am learning something about colour and condition. Then, stare as I would at the two small heaps on the Major's outstretched palms, both looked exactly alike to me. I could only keep on looking with what I hoped would pass for an intelligent and receptive expression, and pray that the Major would not suddenly ask me which sample was which.

Towards the end of the week a watery sun came out to battle with the frost, and we were able to get up a few carrots and some of the sugar beet, though I seemed to spend most of my time writing out carrot lists and tearing them up again. And no sooner did I arrange for lorries and trucks with Barrett's and the Westham station-master, than I would have to ring up and cancel them.

By the middle of the following week, the frost had set in hard again. Kirk, Noller, Bowes and Kettering went down with flu. Jeff and Will overhauled their tractors round by the fitter's shop. Peel and Catton went off to the Hall with the threshing tackle, to thresh out a couple of wheat stacks for Major Rivers. Peacock set off on a round of repairing gates, railings and bullock bins. Patfield and Harry Buxton went from fence to pithole and pithole to fence, gassing rabbits. And the remainder of the men were set to trimming hedges, sawing wood, cutting up hay, carting straw, and all the other odd jobs that are left for bad weather.

The many drains which criss-cross the Manor meadows are cleared out at least twice a year, generally by Isbell, who is, as Marsham says, "a rare worker. But somehow he don't fare to get on with the rest on 'em. Not nohow."

Sighting him from the carrot house one morning, where I had been running through the sack list with Bushell, I waded across the narrow stream above the top pool, and walked down the 'strip' to see how he was getting on.

As I came up he was just finishing the shallow stretch of water which lies between the upper and lower pools, parallel to the river proper. Next he was to start on the dyke which runs along the bottom of Richmond's Place and joins the river just below the Fair Meadow. But what had he better do about that plantation drain on the way? he wanted to know. Was he to stop after it this time? Or would it go till the next? Would I come and have a look?

On our way we had to cross the old bridge which years ago connected two corn and paper mills, and must have carried its share of traffic, but now drowses beside a clump of beech trees whilst any traffic which comes down the Ashacre road is carried by the iron Council bridge put up over the ford on the far side of the pool nearly thirty years ago, and the mills are only a memory.

"Though they do say as you can hear 'em working still, now and agen," volunteered Isbell, as we stood for a moment leaning over the parapet to watch the water surging out from under the bridge, ploughing a wide turbulent furrow across the pool. "Goin' to be a man dead in Silford afore the week's out, an' all."

"What do you mean?"

"You see that foam on yon side of the gush?"

On the Silford side of the swirling water a longish blob of biscuit-coloured froth was held in the eddies.

"Well, what about it?"

"Sign of a death within a week," returned Isbell with melancholy relish. "Never know'd it wrong. Round for a woman, long for a man, an' a small un for a child. An' if it's this side of the gush, it's someone in Ashacre. Now do you see if that ain't right."

There can't be any connection between a blob of foam appearing on Silford Pool one Tuesday morning, and an old man having a fatal stroke at the other end of the parish on Friday afternoon.

It just happened.

⊱ V ⊰

SILFORD, though little more than a hamlet itself, has a number of offspring, still smaller hamlets, within its boundaries. Mrs Heyhoe bicycles to the Manor from 'Nowhere', which is the official postal address of the cottage she occupies with her husband and son on the brink of 'Nowhere Beck', the shallow stream which separates Silford from Banwell Common. The half dozen or so small houses down the lane opposite the carrot washer are known collectively as 'Pockthorpe'. When, some centuries back, the river suddenly changed its course, a fragment of Silford was marooned: and though now to all intents and purposes in Ashacre, on Ordnance Surveys and such it retains its Silford status and rejoices, not inappropriately, in the name of Fustyweed.

The cluster of cottages on the crest of Silford Hill also leads a separate existence. They even have a letter box of their own affixed to the telegraph pole on the main road opposite the end of their lane, with two collections a day. And most of their shopping, like that of many country-dwellers, is done at the back doors of the various trades-men's vans who are carrying on, whether they know it or not, the old tradition of the itinerant pedlars. Whilst Silford Hole has other inhabitants besides the Reeves . . . not quite so tucked away as the Eighth Dwarf and his wife, although actually inhabiting the Hole proper if I read the signs aright. For surely it is the huge scooped-out pit, site of an old lime-kiln, which lies between Three Bridges and the lower half of 'Limekiln' Breck, which has given the Hole its name?

The row of three cottages backing on to the by-road which winds down from the five-cross-ways on the Wendeswell boundary are barely seventy years old and indistinguishable from thousands of similar cottages scattered up and down the countryside, the poor but honest parents of the modern 'council house'. But you can search the country without finding anything resembling that . . . or perhaps I should say those . . . below.

Seen from the road or from the aforementioned row of three from which it is separated by some twenty yards of nondescript garden, only a single-storey dwelling is visible . . . a 'flat' inhabited by Isbell, his wife and two children, and Oby, who is Mrs Isbell's younger brother. But if you approach via the Three Bridge meadow, you see a three-storeyed house set in the hillside, with a delightful façade of squared flint inset with little pointed windows which catch all the mid-day and afternoon sun.

Lichened apple trees to left and right partially conceal the two flights of steps which lead to the upper world. Short thorn fences and a few tall oaks separate it from the Hole on one side and the Three Bridges field on the other, and enclose a garden which produces flowers and vegetables of a quality and abundance unknown in any other garden in the parish, not excluding the Manor's. For the jack of all

trades, Peacock, whose domain it is, has at least mastered
this trade.

Inevitably I should have become aware of Peacock in
time, for he is not one to allow himself to be overlooked in
whatever company he finds himself. But our better acquaintance was expedited by the frost, a patch of ice sending him sprawling just as he was crossing the yard to the
cart-shed to repair some shafts, with a newly sharpened
saw in his hand.

As I bound up the other, which had become involved
with the saw in a highly painful fashion, I offered the sufferer a cup of tea, which I usually included in the treatment.

"Tea? No, thank ye, miss. I'm glad to say there's none
passed my lips these twenty-five year an' more."

He was, it appeared, a connoisseur of wines, all of which
he made himself. Sugar beet, mangolds (cattle beets),
barley, sloes, bullaces, potatoes, nettle, elderberry, rhubarb
. . . there wasn't nothing you could tell him about any of
'em that he didn't know. But the king of 'em all was the
orange. Any fruit merchant would let you have a case of
rotting oranges for next to nothing. And you couldn't beat
'em. He'd got a bottle on his bike now, just the right age. I
ought to try it. And he was off up the yard to the cart-shed
and back again before I had finished repacking the first-aid
box.

Over the orange wine, which was surprisingly good (not
to say potent) I heard the first instalment of that entrancing
serial story, 'The Life and Times of John Herbert Peacock'.

He had worked on the Manor farm for nigh on twelve
year now, he informed me . . . a length of time which,

though nothing in comparison with that of several of his fellow labourers, had never ceased to astonish him, for it was the only place where he had stayed more than a few months.

Running away from home (a cottage on the far side of Banwell Common) at fourteen, he had been at times ship's cook, farm labourer, miner, fiddler (when he had earned many a meal and shilling entertaining the company in various hostelries and still did on occasions), house-painter, car-penter, and a dozen other things besides. He might ha' done something with hisself, for he could turn his hand to anything, he said pensively, excepting that somehow he always managed to leave something.

He does, indeed.

If told off to repaint, in those brilliant scarlet or corn-flower blues which always make farm implements look so cheerful, the wheels of some tumbril or wagon, he invariably misses out a spoke. If it is the window frame of a cottage, he never remembers to unfasten the window, and always has to be sent back to finish the unpainted fragment revealed when the window is opened. And there was one epic day shortly after my arrival when, having been told to nail up a certain gateway between the Fair Meadow and some newly sown wheat (since trespassing lovers persisted in leaving it open and the cows had already been over the wheat twice), he ignored the scheduled opening altogether and instead nailed up the only exit on to the road with such efficiency that the gate had to be broken before the cows could be released and driven home for milking. As Major Rivers remarked with considerable bitterness on this occa-

sion: "The little devil knows his job all right. But it's absolutely useless to leave him alone. You mustn't take your
eyes off him, or he's bound to bugger it up."

The only thing over which he could never go wrong was
his garden. Like 'me mother', he had green fingers, he told
me. If you hadn't got 'em you might just as well lay off gardening for all the good you'd do. Like Ernie. He'd never
make a gardener if he did nothing else for a month of Sundays. He'd better by half stick to his stables.

'Ernie' is Croft.

"The best he ever done was afore your time, miss. About
five year ago. Ernie, he was laid up with the 'flu, or something. So Mr Nicholas, he put me on gardening, and I grew
a crop of onions such as you don't often see, and about twice
as big as any Ernie's ever managed to coax up. You should
ha' seen the mucky little warmints he grew the year arter.
I'd got some solers, same as I always have. So one morning
I browt a string of 'em up and gav' 'em to Mr Nicholas in
the yard. In a day or two, along comes Ernie. 'Got something to show you, John,' says he. And takes me along to
the gig-house. 'Just you look at them there,' he says. 'Tha's
the way to grow onions, John,' he says. An' shows me
m'own onions hanging on the wall!"

"What did he say when you told him where they'd come
from?"

"Told him where they come from! Wha's the use a' telling *him* anything? He'll never be a gardener," replied Peacock, with withering scorn. "So I never said nothin'. Not
to him. I mentioned it to one or two of 'em in the yard,
though. So I 'spect it got round. M'own onions!"

Onions are not the only vegetable Peacock grows for winter storage.

"Why don't you ask Peacock if he's got a million to spare?" suggested Mrs Heyhoe helpfully, a couple of mornings later when I was grappling with the menu. "Then you could try a million pie."

"A million?"

"Pumpkins, some folks call 'em," Mrs Heyhoe translated placidly. "You stew it a bit first. Then you bake it in a pie with raisins, and sugar, or salt and pepper an' chopped up pork. Depends on whether you want it sweet or savoury."

The frost went on.

Marsham's round face lengthened every time he looked down at the unyielding earth. Major Rivers gave up glancing at the thermometer in the porch. Everyone was busy, but there was an exasperated feeling abroad that time was being wasted on non-essentials while the real work was getting well behind.

One evening I was listening to the wireless in the flat when, above the nine o'clock news of mass air-raids on Finnish towns, a slight easing of the tension in Holland and Belgium (where all military leave had just been cancelled), and more about food rationing, a high-pitched affronted miaow made itself heard.

Only one creature in this world has a voice of that calibre.

Somehow or other, Master Samuel Pepys . . . who has reached an age when he prefers to spend his nights indoors, though there are times when he behaves like a three-month-old kitten . . . had contrived to get himself locked out, and was now telling the universe about it from a seat on the scullery window-sill, immediately below mine.

As I opened the side-door to let him in, an indignant grey ball catapulted past me into the hall and made for the stairs, though not before I had caught a glimpse of damp-flattened fur.

I stepped out on to the gravelled path.

When I had blacked-out, the night had been cold, sharp, clear-cut. Now its edges were blurred. The stars which filled the sky seemed oddly remote, unreal, withdrawn. There was a change, indefinable but unmistakable, in the atmosphere. And what I took to be a fine misty rain was falling.

If a thaw was setting in at last, it would alter all tomorrow's orders. We could start carroting again. Patfield must down gas-pumps, and round up the women first thing. I must arrange for trucks and lorries. Then there was that sugar beet. I must ring up the Major directly Marsham came in the morning. I was so pleased with myself for thinking of all this before I was back in the flat that it was all I could do to refrain from dialling Ashacre 57 right away.

Fortunately I confined myself to rubbing down Master Pepys with the kitchen towel. For when I woke next morning my room, though still dark, appeared several degrees lighter than usual. The various 'noises off' seemed curiously subdued. There was a new quality in the air. Whilst I was wondering sleepily what it was all about, Marsham's voice announced plangently below that the old women hadn't half bin plucking their geese during the night.

"There's a fut a' snow on the ground."

"Snow!" I leapt out of bed and reached for my dressing
gown.

Sure enough, there it was . . . an all-enveloping mantle
spread over roofs and walls, weighting down trees, trailing
its folds over fence and field, turning the prosaic workaday
world into a fairy land of dazzling white. I stared en-
tranced.

"You'd better give the Guv'nor a ring." Marsham's voice
broke the spell. "Not but what there's nothing he's likely to
say as I don't know afore. The wind's drifted the mucky
stuff. So the first thing we'll ha' to do is to dig ourselves out.
It's over the tops a' the hedges in the church lane, and eight
feet deep in the Ashacre road. If the Guv'nor wants to call
in this morning, reckon he'll have to come on Shank's
pony."

At the other end of the telephone Major Rivers confirmed this. Ashacre Hall was cut off as completely as the
Manor.

"I might be able to get through this afternoon, but you'd
better not count on it. Still you can't go wrong. Peacock can
keep on carpentering and Patfield gassing . . . this wind
must have blown plenty of the holes clear. Vincent and
Mark can find themselves a job in the fitter's shop. There
are several things want doing to the ploughs and the Cambridge rolls. And everyone else will have to go clearing
snow."

Freed from the necessity of waiting for Major Rivers or
the post . . . for no postman, certainly not the Ashacre
worthy who normally follows his calling with the ardour

of a dormouse and the reckless abandon of a competitor in a slow bicycle race, was likely to negotiate the trackless waste just yet, even if the General Post Office van had been able to force its way through from Norwich . . . and if Mrs Heyhoe should turn up, she could get on by herself, I lit the flat fire, breakfasted, and clad in rubber boots, mackintosh and practically every woolly garment I possessed, set out to inspect all the cattle. What if I had been round the lot only yesterday, examining each one with anxious care to make sure that it was conforming to the rules laid down by Major Rivers for 'good doers' . . . a well curved stomach, front legs wide apart, a tendency to lick itself, and, if lying down, a cud-chewer? Any excuse was good enough to abandon cow records and carrot statistics, and escape from the office into this exhilarating brave new world.

As Marsham had reported earlier, the lane to Silford village was filled fence high. But I had no business in that direction. My way lay along the 'loke' running up from the yard, past the bottom of the drive, to the Ashacre road: and this was only buried below a mere twelve inches or so. And opposite its farther end somebody had failed to shut the gate in the 'Hunts and Gooch' fence, which had resulted in something like a gap in the high white wall which had once been Ashacre road. So that I was able to scramble through into the field beyond and then climb the slopes of Crabb's Castle.

By the time I reached the top, the wind had died down to an intermittent whisper, and the sun had come out, illuminating the cold purity of the snow with an almost unbearable splendour. Not since one never-to-be-forgotten Christ-

mas holidays spent in an Essex village as a child, had I seen
so much massed beauty. In London snow's loveliness is an
affair of minutes, seconds even. Often it has degenerated
into mud and slush before you know that it has fallen. Even
in the parks it is quickly blotched with soot and grime. It
was good to be alive in Silford on a day like this, even
though there was a war on. I wanted to dance and sing.
Since there was no one to see or hear me, regrettably I did
both before squeezing through a gap into 'Mautby's Hall',
and so to the Silford Hill buildings where dwelt the forty-
seven head of cattle tended by 'Old Tom' Buxton.

"What a morning, Tom."

"You may well say what a morning, Miss. An' more
a'coming if you ask me. An' what good do it do? Tha's
what I want to know. What good do it do?"

Old Tom, it needed no Dr Thorndyke to deduce, was no
sharer of my admiration for snow. Nor was anyone else on
the farm apparently. Although Marsham was moved to relate the story of other and greater snows as the day wore on.

"It must be fifty year ago . . . No, it warn't. It 'ud be
more like fifty-two. Wrong agen. It was fifty-one. Only
the roads was blocked from east to west that time. The
turnpike was worse than what the Ashacre road is now.
That was when my brother was a boy, afore he went to
Australia; and they were all playing about making houses
in the snow up by the 'Hart'. I've often heard my fayther
laugh about it."

"But what were you doing? Why weren't you making
houses, too?" I enquired thoughtlessly.

Before I came to Silford, I had prided myself on being a

fairly good judge of ages, and would have backed myself to place almost anyone within a year or two. But the agricultural labourer grows old in a manner all his own. In some mysterious way he contrives to leap from boyhood into early middle age at a single bound, and then stays there unchanged as far as appearances are concerned, for the next half century. At the Manor only Oby provides the exception that proves the rule, his bright pink face, wide open eyes, and half open mouth suggesting perpetual sixteen. The rest might almost be any age between thirty and seventy. Marsham, with his shapeless old felt hat crammed well down on his ginger head (I have never yet seen him without it, even when I have called in at his cottage with a change of orders. . . . I believe he sleeps in it), looks as ageless as anyone else. But I had noticed grey hairs protruding from his son Roger's cloth cap, and had deduced from this and other clues that Marsham must have reached the later fifties. Apparently I had made a bad gaff.

"Why I wasn't hardly born then. I wasn't more'n two year and nine months when me brother went away. I remember another lot though. That'd be more like forty year ago, when me sister, Mrs Heyhoe that now is, come along.

"We was living at Morley, then, and I had to go and fetch the doctor from Meddenham. The first four mile might ha' bin worse. Then, keeping close to where the fence ought to ha' bin, because I thought that wor the safest, I went in to a holl right up to me neck. I thought I should never get out agen. An' I was soaked to the skin, along a' having slipped into several more, afore I got to the doctor's.

"An' then he said he warn't acomin'. How was he going

to get a hoss an' cart the best part a' ten mile with the roads same as they were? Well, I couldn't help that. And in the end he came on hoss-back. An' danged if he didn't turn the corner too sharp, an' he an' the hoss got stuck in the holl same as I did. Alice had been hollering the place down an hour an' more by the time he got to me mother."

As day succeeded day my own enthusiasm for snow began to suffer a sea-change.

Egged on by Mrs Heyhoe, who is as full of country remedies as a mole's coat is of fleas, I ran solemnly round and round the tennis-lawn, barefooted, to cure my chilblains, which this drastic treatment certainly did. But otherwise I was soon inclined to echo 'old Tom's' query of 'What good do it do?'

The sun never reappeared after those first few brilliant hours. Nothing relieved the deadly sameness, the depressing monotony of the leaden grey skies and that unbroken vista of white. It ceased to be an amusing venture to wade kneedeep through snow. To find each morning that more had fallen, or that the wind had playfully blown earlier falls back into the narrow one-way streets which were being so laboriously carved through the blocked roads. On the rare occasions when I ventured out in Amy, I had to be accompanied by Croft and a pick and shovel, which we never failed to need. A sortie into Meddenham for spare parts urgently needed by Mark for repairs, was achieved only after many days and an incredible mileage, since all the more direct routes were still snowed up. Major Rivers skidded into a solid wall of snow on the way to Norwich and turned completely over, though fortunately without damage to himself and only minor injuries to the car. Telephones were continually and inconveniently out of order. Pigeons descended in clouds on the kale and brussels sprouts. Flocks of birds of all sorts and sizes fought pitched battles for crumbs on the kitchen window-sill. Rabbits missed by Patfield and made desperate by hunger, gnawed everything in the vegetable garden and even rashly invaded the front-door step, to the delirious joy of Master Pepys. Everywhere everyone talked snow, cursed snow, ate and drank with snow, waked and slept with it.

Everyone? No, not quite.

"Snow?" said Croft one afternoon as we groused together.

"Don't you never listen to the wireless, or look at the papers, miss? There ain't no snow. It's only a rumour. Do you see, they'll be giving us pictures a' splashing in the sea at sunny Brighton afore the week's out. As though Jerry didn't know what the weather's like over here's well as we do, with old Haw-Haw always telling him. I hope they catch a cold same as the one I can feel comin' on. That'd learn 'em."

Croft was not the only sufferer from the weather. Except for Noller, who had come back days before he had any business to be out of bed, and was now crawling about the place like a willing corpse, the original 'flu casualties were still away. Marsham sniffed and sneezed cheerfully all over the office while Mrs Heyhoe staged similar exhibitions in the kitchen, both of them impervious to hints as to the wisdom of a day at home, though they accepted the aspirin and quinine I offered in self-defence. Soon Bushell, after announcing at intervals for a couple of days that he 'felt queer' — he certainly looked it — disappeared off the strength: to be followed in turn by Isbell, Luke, Oby, Hill and then Lambert.

"Reckon Peachey'll be the next," said Marsham with a broad grin, as we filled up the labour book that evening, and discussed the question of whether Catton or Ted Long, the only labourers able to milk, should be sent into the cowhouse in Lambert's place.

"Why Peachey? He was all right this afternoon."

"It's his missus," confided Marsham, looking more like the Cheshire cat every second. "She's allus round to Harry's or he's round there, when Peachey ain't about. There've been some rare old how-do-ye-dos, what with one thing and another. Harry gets longer for dinner than what Peachey has, being a cowman. And gets back late on top a' that, 'cause she sends him down to Ashacre to do her bit a' shopping. Lily does her best for her feyther. But she's married, and got her own home to see after, down to Banwell Common. And when she spok' to an old sweetheart a' Harry's, what would ha' come and kep' house for him, Peachey's missus, she soon put a stop to that. So there y'are."

I knew Mrs Peachey by sight, as one of the carrot pickers. And occasionally she came into the yard to fetch milk. A smallish woman with the thickened figure of the average country-woman, an untidy frizz of straw-coloured hair straggling from under an emerald knitted cap, opaque blue eyes set slightly too close together in a sallow face which would have been none the worse for some soap and water, and, almost invariably, a cigarette dangling from one corner of her small thin-lipped mouth. An unlikely subject for a *grande passion,* I should have said. Though not so unlikely as Lambert with his purpling nose, hang-dog air, and that all-pervading aroma of cow. Well! Well.

"You know, miss, it ain't right." Marsham had also been in labour with great thoughts, though the sentiments of which he was now delivered were a little unexpected, coming on top of that opening sentence. "Time is time. An' time should be kept. Making a start at twenty-five after six! An' half-an-hour late back from dinner's nothing. You want to keep his time for a week when he comes back. The Guv'-nor'll have to be spok' to about him if he don't alter. It only upsets the others. And it ain't the thing at all."

It wasn't, I agreed.

"Pity. Cause he's a rare hand with cows when he likes. There's old Jenny. Kick the cowhouse to bits as soon as look at you. They'll have to rope her to milk whiles Harry's laid up. No one else can't do nothing with her. But she's as quiet as an old ewe along a' him. An' when he goes to fetch the cows up, she'll come running acrost the medder, blaring away, the minute he gives 'em a call. Well, I must be getting a move on, do I shan't catch Catton afore he leaves off. Goodnight, miss."

"Goodnight."

Snow, more snow, and yet more snow drifted down from that relentless sky.

Major Rivers had it from a friend who had a friend who knew someone who had a friend at the Foreign Office, that if this weather went on for another month . . . and there seemed to be every prospect of it doing so . . . it would just about break the back of the war. So we all dwelt at length on this comforting theory, and Marsham refrained from muttering more than twice in any one day about the work that was piling up. Fences were trimmed almost to extinction. Repairs were carried out until there was nothing left to repair. Joe Gidney and George Long sawed wood in the barn until the engine's piston rings gave way. Cecil Peachey ranged over the seed kale and savoy cabbage fields in an attempt to intimidate the everlasting flocks of pigeons, which in spite of the vigilance of Myhill and the 'clock-gun', which was set to fire a shot every hour, had already spoilt the four acres of brussels sprouts at the bottom of Crabb's Castle, and were growing more voracious each day. Muck was carted out of every yard and spread on the snow-bound fields. The last of the corn was threshed.

"Leeds rang me up last night," said Major Rivers one morning at the beginning of February. "About savoys. Normally we shouldn't touch them for another seven or eight weeks. But other people are sending some in. And they're making quite a price, it seems. Nothing else to be had. You'd better get on to the Westham stationmaster right away. It's no good trying Barrett's. The roads are still too bad. But if you can get trucks at Westham, you can start cutting on Swains tomorrow. Patfield can tell you how you're on for cabbage bags. There should be plenty in the sack-house. And you can drop a line to everyone to say you'll be loading savoys and will they send on more bags at once. Where's the carrot book?"

Though it goes by the name of the one vegetable, the carrot book is the imposing ledger in which all sales of market garden produce are recorded. Fishing it out of the drawer it

occupies in solitary state, I put it on the desk in front of the
Major who hastily began turning over its multitudinous
pages.

"Spring cabbage . . . turnips . . . peas . . . savoys. Here
we are. You can get a rough idea from this how many each
man can do with. It works more or less in the same way as
with carrots. You have to watch them for prices, and slow
down on deliveries to those who're paying the least."

We are not really popular with the Westham station-
master, who can never understand why we cannot always
give him a week's notice about trucks. If he would spend a
week on a farm he might find out. However, for once he
made no difficulties, so Patfield spent the afternoon sorting
out sacks and I finally achieved a reasonable cabbage list.

On Swains there was still four or five inches of snow, and
most of the savoys were invisible. But unlike the carrots,
they were above ground and could be scooped out. By the
time I reached the field, four men were busily cutting, another ten were shaking the savoys clear of snow, gathering
them into heaps and packing sixty to seventy pounds of
them into the cabbage bags supplied by most of the market
salesmen, or the nets favoured by Messrs Leeds and Bur-
ford. While George Long, armed with a sack needle, sewed
up the full bags with red binder twine, ready to be carted off
to Westham by Mark. An icy wind was sweeping over
Swains, which stands high as Silford levels go.

"They've been cutting savoys at Wendeswell Hall this
last week or so. And Mr Blyth, he's giving his men glove
money, so one on 'em was telling me," George Long hinted
delicately on Thursday afternoon. "You might just mention it to the Guv'nor."

I mentioned it.

"First I've heard of it!" snorted the Major. "Doesn't sound like old Blyth, either. Glove money, indeed. They'll be wanting Sunday suits next. And you can tell George I said so."

It would pay us to get carrots up dirty, if only a few, wrote Messrs Prestwick and Marriott, Root and Hopkins, Mitford and Sellars. "Load carrots. Urgent," telegraphed A. L. Gardiner, and E. H. Oliver and Sons. "Dirty carrots are making up to a pound a bag," R. A. Leeds rang up to say. "Mr Holmes, out yon side a' Meddenham, he's using one a' them there gyrotillers and blasting the carrots out a' the ground, so I've been hearing," Marsham reported to me.

A day or two later the merest suspicion of a thaw set in. Patches of brown earth began to show here and there in the fields. The snow walls beside the roads looked the least bit lower.

Peachey, (who had had, as Marsham prophesied he would, a mild attack of 'flu, but had come back to work again the day after Lambert) ploughed out an acre or so of carrots on Booters, and after a certain amount of negotiation, the men agreed to pick them on piecework at half-a-crown a bag. Wet, dripping, muddy, they were loaded on to the small trailer by Mark and Burton (the roads were too bad for the five-tonner) and carted off to Westham. Within forty-eight hours of the first load's departure, frantic telegrams were pouring in from London, imploring us to stop loading. Carrots unsaleable unless washed. I reached Booters just in time to stop another load being carted off, and had to send Mark and Gidney back to Westham station to retrieve the two-and-a-half tons they had already carted in that morning. Next day they were washed and sent off again, several half bags disappearing in some mysterious way between Westham and Bishopsgate, several more arriving (according to the salesmen) in an unmarketable condition, and most of the remainder making a price that made Major Rivers look glum, and doubt the wisdom of getting any more up.

This particular problem, however, was settled hardly before he voiced it, by the temperature, which had once more fallen several degrees below freezing point, and stayed there. More snow fell.

Hurrying back to the flat fire and lunch one day, I was surprised to find half-a-dozen small children loitering by the front door. As they saw me, they burst into a shrill chant, which they repeated over and over again. The words were undistinguishable.

"Goodmorrerwalentine," volunteered the largest, a mature maiden of nine or so, when I asked what it was all about.

The rest gazed expectantly.

I gaped back at them. Then, murmuring hastily "I'll be back in a second," I bolted for Mrs Heyhoe and an explanation.

"Tha's right. It's St. Walentine's day. Time was when there'd be two or three score on 'em going round from house to house. Mrs Irstead and Miss Kate, they used to throw 'em oranges and apples and pennies to scramble for. But they don't go round so much nowadays. Quite right, too. A proper nuisance they used to be. Now don't you go giving 'em too many a' them biscuits, miss. Do you'll have half the village up here this afternoon. And make 'em scramble for the pennies. Tha's the proper way."

Twelve small paws delved recklessly in the snow for coppers.

Six small mouths performed the double feat of champing biscuits and yodelling "Good morrow, Walentine," as their owners made off down the drive.

But nobody, not even that perambulating encyclopaedia of folk-lore, Mrs Heyhoe, could tell me why the anniversary of St. Valentine's death at the hands of the Romans should be remembered in this way in a Norfolk village: or whether or no it has any more connection with the saint than the custom of sending lovers' greetings which has grown out of the old belief that on the fourteenth of February birds begin to choose their mates.

All over Europe the frost held and icy numbing winds went skirling. They blew across England, and ranged unceasingly over Silford and the cabbage fields.

Savoys fetched fabulous prices, though never quite as fabulous as those appearing under 'Market Reports' in the local press, which George Long wistfully quoted to me at intervals. In the end he hypnotised me into mentioning glove money again.

"It's an imposition!" the Major fumed. "Soon there won't be a single job they'll do for their ordinary wages. Can't say I blame them, though. This weather's an absolute b——. Find out from George what old Blyth is giving his chaps. And tell 'em they shall have something to share out when the savoys are finished, if they put their backs into it."

≫ VI ≪

SLOWLY but surely the brown patches on the fields grew larger. Trees and hedges stretched their cramped and stiffened limbs. Water coursed down the roads between the dwindling banks of snow. The river rose and flooded all the lower meadows. From the Church Lane, which we had never had time to dig out, a miniature torrent raged down the hill and through the yard into the horsepond.

In hollows and under north fences snow was to lie about for weeks yet 'waiting for more', as 'old Tom' informed me, correctly enough. The frost made unexpected sorties, set an ambush here and there, sent out a number of raiding parties, re-took the old position. But winter's main attack was spent. Inch by inch, the advancing spring drove it back. On the rockery near the top of the drive, the melting snow revealed a primrose plant of which there had been no sign when the first flakes fell. Snowdrops and daffodils pushed up pale green shoots through the sodden earth. The wind, still high and keen, began to dry the land as well as the linen flapping up and down on the lines which criss-cross most of Silford's back-gardens. Came the day when Vincent took the 'crawler' up to Ram's Breck and Leslie Heyhoe set off for Three Bridges with the Case tractor, to start ploughing again. Peachey, Hill and Ted Long went harrowing with a pair of horses apiece. Roger Marsham with three more took the double furrow plough across Crabb's Castle, whilst Heyhoe with another pair and a single furrow dealt with the hollows on Prisoner's Close and Broomhills.

"How we shall ever ketch up, I dunno," proclaimed Marsham. "Nawthen'll get done as it shud. Tha's a sure moral. There ain't the time, with March on top a' us already. We shall just hev to do the best we can."

The last of the sugar beet was lifted from the slippery water-logged Sixteen Acres and sent in to the factory, some on Barrett's lorries, some by rail from Westham, one of the latter trucks being returned as useless though its fellow, lifted and loaded at the same time, was accepted without comment.

Acrimonious words passed over the telephone between the Westham station-master and Major Rivers, Major Rivers and the factory, but nothing could be done.

"It's funny stuff at that," the Major said thoughtfully, as he finally replaced the receiver. "Low sugar content, half Silford sticking to it, and the frost into it properly. I've never seen anything like it before. You can think yourselves lucky that you didn't get the lot shot back on to you. There's a poor little devil of a small-holder at Dennington who got all his up by Christmas, and the contractor never carted it in until last week. They refused to take so much as an ounce at the factory. And there isn't a thing he can do about it. Can't go for the contractor. No insurance. Nothing. Just rotten bad luck. One of the joys of farming, Miss

Harland. And not the only one. You're going to catch a
cold over the carrots as well, I shouldn't wonder."

We had taken several walks over the carrot fields lately:
Hunts and Gooch, Pound Close, seven-eighths of Booters
. . . about fifty-five acres in all. At a conservative estimate,
there should have been some thousand tons of carrots
moulded up in those neat ridges which ran rank on rank
across the three fields.

Should have been. But something had happened to the
carrots.

That innocent little thaw in early February had allowed
water to seep through into the soil around and below them.
Then the frost had set in again, striking deep into the
ground. Now that it was at last relaxing its grip, the full
extent of the damage was becoming apparent. Here and
there a few hardy carrots were putting out one or two fresh
shoots. Given time, they would recover. A few more might
follow their example. But the bulk of the crop was frosted
beyond hope, was rotting as it stood. As the days lengthened, they simply melted into the ground. A gang of men
on piecework combed Hunts and Gooch for survivors, retrieving an average of half a ton per acre, which they rattled
over the potato riddler to knock off some of the mould preparatory to weighing them up in half bags to be sent dirty
to London. But the Major stopped them before they were
ready to begin on Pound Close. Even though they were
making up to twelve shillings a half bag, it was a waste of
the men's time. There was nothing for it but to plough the
carrots in.

"Vincent can move into Hunts and Gooch directly he

comes off the Upper Common. And then on to Pound
Close, tell him. Booters will have to be left until it dries off
a bit. I don't suppose you'll be able to touch that for another three weeks at least. You'll have to watch your opportunity. And the first wet day Bushell can have some help
and give the carrot house a thorough clean up."

We finished cutting the four acres of savoys at the bottom of Gunspear, averaging two hundred and fourteen
bags an acre. In the lucid intervals between showers of rain
and sleet, flurries of snow, and sudden returns of frost,
Roger Marsham drilled oats on Easter Field and Fair Close,
barley on Sidegate Breck, Ram's Breck, and Three Bridges.
Kirk and Meale sowed manure, the latter scattering nitro-
chalk by hand on the winter wheat, the former sowing basic
slag (for which we should presently draw a bonus from the
Land Fertility Commission), three hundredweight per
acre on all the barley land, with the manure drill. The last
of the convalescents drifted back to work, though the roll
was still incomplete, the latest absentee being Myhill who,
bicycling over the five-crossways by Ashacre school on
Easter Monday, his mind on other things, had come into
conflict with the butcher's van and was now at home suffering from what the doctor diagnosed as shock, and Marsham
uncharitably and rather unjustifiably called something else.

"Why the go-to-flying don't he look where he's going?
Parfitt ain't half bin mobbing about the dents in his mud-
guard. He reckons Myhill must ha' bin in the Fox and
Hounds. No more it wouldn't surprise me if he hed an' all.
Not but what I was in there meself the night afore last,
along a' me brother-in-law. An' who should walk in while

we was there but young Gordon Howard, him what used to
have Oby's job afore he went for a sailor. Seems he was in
that do they had a few weeks back when they took them
men a' ours off that there Altmark. A proper go, it was, Gordon said. He used his fists like the rest of 'em. Knocked their
heads together and cuffed 'em overboard right and left.
Tha's the stuff. I'll lay the ones he set about don't forget
Gordon in a hurry, for all he's quiet as they make 'em, do
he don't get upset."

Before the end of the month Captain Irstead arrived on
seven days' leave.

Slight, dark, and something under middle height, he was
extraordinarily like his sister to look at. But there the re-
semblance ended. He had none of her birdlike quickness of
mind and body. Where she swept along on a tide of resist-
less energy, he hung back, or moved lethargic, his speech
matching his gait. And in contrast to her direct frank
friendliness, he was reserved almost to the point of sullen-
ness.

Ostensibly domiciled at the Hall, he spent a certain
amount of time there and visiting friends, but more prowling moodily over the Manor fields and unlocking and lock-
ing doors inside the house, eyeing me suspiciously the while
as though, like Mrs Dack, he wondered how far I was to be
trusted with his precious house and farm.

Since the trap in which he was caught was none of my
setting I could have wished that he had been a little more
affable in it, though I could understand how he felt.

Mrs Rivers had told me of his deep rooted love for the
Manor, every tree and stone, every meadow and field, every

brick and tile of it. The place was a part of him, like an arm
or a leg. But while he had always been ready to spring to its
and his country's defence, and for the last five years had
gone regularly to 'camps', on T.E.W.T.s, and addressed recruiting meetings, I don't think he had ever really bar-
gained for war. He had certainly never envisaged any
lengthy separation from Silford and all that it stood for, or
realised that it could go on without him.

And then the unimaginable had happened.

The arm had been amputated, and miraculously surviving the operation, had blossomed into a life of its own. It
was the parent body which failed to recover.

The men greeted him cheerfully, asked how he was getting on with the war, what he thought of the Finnish peace,
and how he liked France. But they turned automatically to
the Major or me if it was a question of anything to do with
the farm, talking across its rightful master as though he
ceased to exist. It was hardly surprising that he should be
somewhat disgruntled, and find a good deal to dislike about
life at the moment. Or perhaps it was merely that he dis-
liked me, a compliment that I was in a fair way to returning
with interest, since my own outlook on life was getting
more and more jaundiced every second.

For the last few days I had been feeling thoroughly edgy
and irritable, with a tendency to headaches and recurrent
attacks of self-pity. At dinner at Ashacre Hall on Captain
Irstead's last night of leave, the food, well chosen and ad-
mirably cooked as meals at the Hall always were, tasted
like wood. The voices of the young Rivers, normally bear-
able, seemed to have acquired new and ear-splitting quali-

ties. Bridge had ceased to be an amusing pastime and had become a penance. When I got back to the flat, I took my temperature before getting into bed. It was just below 103°.

I am an ungrateful patient, I know.

Not for me the soft ministering hand soothing the fevered brow or smoothing the tumbled pillow, if I can help it. The less I see of my fellow creatures on these fortunately rare occasions, the better for all concerned.

If I could have had my way, I should have barred my flat against all comers save Master Pepys, crawling out to fetch such necessities as milk at intervals, and otherwise staying in bed, where I could refill my hot water bottle and brew myself innumerable cups of tea with the help of the electric kettle beside my bed, without worrying anybody.

But this, alas! was out of the question.

I could and did refuse Mrs Rivers' instant offer of transport and the spareroom at the Hall, and suggest that she should confine her calls to the telephone because of the children. But nothing could keep out Mrs Heyhoe.

"Lor, bless you, miss. I'm used ter looking after illness. Send for Alice, my folks, they allus say when any on 'em ail anything. And I don't never take no harm meself," she cried, when I tried to scare her off. "Now just you lie back an' take it easy, and you'll soon be as right as rain. What could you fancy for breakfast, now? I could soon have you up some bacon and eggs."

"I don't want anything at all, thank you. I've had a cup of tea and some aspirin. I'm going to try to get to sleep."

"But what about your dinner, miss? You'll be wanting something solid by half arter twelve."

"No, I shan't. If I have anything, it'll be another cup of tea."

"You'll find it won't sit comfortable, miss. Not without something to settle it," said Mrs Heyhoe disapprovingly. "You put me in mind a' my sister-in-law. A rare trouble she was to get to take anything. A' course she had to be careful what she et. She'd got the diabetes, and I had to keep injecting her with that there insolent. But she ought to ha' et what she could. I shall allus hold as she'd ha' bin alive today if she'd eaten heartier."

"I don't want anything except a cup of tea, thank you. And now I'm going to sleep."

I closed my eyes, turned my face to the wall, and a moment later had the satisfaction of hearing Mrs Heyhoe tiptoeing noisily out.

The telephone rang twice within the next three-quarters of an hour, once to inform me that Oliver and Speedwell had cleared yesterday's savoys at thirteen stroke six per bag, the second time to enquire if I was Mrs Copplestone? No, I wasn't, I answered disagreeably and put back the receiver to find Mrs Heyhoe knocking on the bedroom door.

"Seein' as you was awake, I thought p'raps you could do with a nice cup of cocoa."

The very thought of it turns my stomach, but I hesitate to say so in the face of that trustful smile.

"Feed a cold an' starve a fever," observes Mrs Heyhoe as she reverently places the cup of cocoa on the chest beside my bed. "I don't rightly know which you'd call the flu. Still if you eat something, you'll be on the right side." And is there anything else I can think on now she's up?

"No, thank you. I'm going to sleep."

"I should if I were you, miss," Mrs Heyhoe agrees sympa-
thetically and after no more than three additional queries
as to am I sure there's nothing I want, she takes herself out
of the flat.

For perhaps half-an-hour I am left to hate everybody in
peace. Then Mrs Heyhoe comes marching upstairs again.

"There's Arthurton's man, miss. I've ordered two loaves
for the weekend. Was there anything you'll be wanting
him to bring in the grocery line when he come a' Tuesday?"

"No, thank you."

"Some tinned fruit, now? I always think that slips down
a treat when you're not feeling up to much."

"No, thank you."

"Or a tin of salmon, what I could turn into fish balls or a
pie?"

"No, thank you. Nothing at all."

"I'll make you something with an egg then, miss. You
must ha' something. And you han't drunk your cocoa,"
says Mrs Heyhoe reproachfully. "I'll ha' to take it down
again. It's gone quite cold."

At least she would have to go home at the end of the
morning, to attend to her own affairs. About an hour to go.
I could just about manage it.

"Here's your newspaper, miss, what Drayton's just left."

"I've scrambled you an egg, miss, and made a custard."

"Here's your tea, miss. And I've baked one a' them
sponge cakes you always likes." . . . For Mrs Heyhoe has
no intention of deserting me just yet, but has bicycled back
from Nowhere directly she has seen her own family setting

down to their meal, bless her. It is all I can do to persuade
her to go home to sleep. She doesn't half like leaving me,
she protests. And am I sure I won't be nervous?

No, I shan't be nervous.

"A' course, you've got the telephone beside you. They're
handy things in case a' illness, though I'm a bit shy a' talking down one on 'em meself. I suppose it don't worry you,
though, do it? That's what comes a' having an eddication.
It makes a difference, don't it? Now I'd better be after your
supper, han't I?"

"I don't want any supper," I cry. But Mrs Heyhoe has
already gone. And undeterred by the fact that she had removed the previous trays untouched, ten minutes later she
is back again with yet more food, which she fidgets around
with practically under my nose until in sheer desperation I
sit up and eat something.

"That's better," approves Mrs Heyhoe as I gulp it down.
"You're looking more like yourself already, miss." (Then
either my looks are no indication of my feelings, or my
normal appearance must be worse than I had supposed.)
"Though the flu's a master rum'un," continues Mrs Heyhoe, after much deliberation. "You may try this an' take
that. But it goes on its way regardless. Fares like there's
nothing really do it a mite a' good, like the huping corf.
But you've heard that tale afore, I doubt?"

Even if I have, there is a gleam in Mrs Heyhoe's eye
which tells me that I am about to hear it again. Though actually it is new to me.

"A' course, that was back in me feyther an' grandfer's
time. And farther back nor that. They don't go in for them

sort a' things quite so much nowadays. Though there's a lot in some a' them old remedies for all that. I allus use a cobweb to staunch blood, meself. While there's nothing like a goat's bean wrapped round your throat, for mumps. An' though it hain't took it quite away me lumbago hain't been nothing nigh so bad since I took to wearing a skein a' green silk round me middle. An' when I was a youngster an' kept getting rits (warts) on me hands, many's the time I've stole out at night an' sang 'Oak tree, Oak tree. Please buy this rit of me' nine nights running without letting on to a soul, an' I've never know it to fail. An' once when I hed . . ."

Ordinarily I should have encouraged her, but tonight I would have bartered the most fascinating stories in the world for a few quiet wordless hours alone.

"You were going to tell me something about a cure for whooping cough," I interrupted with feeble determination.

"Oh, ar. So I wor. Well, you see, properly speaking, there ain't none. Do there's no harm in trying. An' Mother, she served us that way when we had a dose on it as young 'uns. Though it didn't make no manner a' difference to the way we wor hawking an' tizacking about, no more than grandfeyther he said it wouldn't. An' he knowed all about it, he said, along a' what happened to old Jimmy Duffield what used to live on Banwell Common. Do it may ha' happened to someone else an' all, but it were like this here, the way grandfeyther he allus told it to us when we was nippers," and Mrs Heyhoe's accent insensibly broadened.

"Jimmy, his little cheild hed got the huping corf, an' he din't know what to do, not nohow. He went to the doctor's an' got har a bottle a' med'cin. But that din't do har a mite

a' good. She just kept a-corfing, an' Jimmy, he was in a
proper state, when he happened on a neighbour a' his what
was also a family man, an' he say to him, 'Jack, my littul
cheild, she ha' got the huping cough. I ha' bin ter the doctor's an' got har a bottle a' med'cin. But that didn't do har a
mite a' good.'

" 'No more it ont, neither,' Jack he says. 'I'll tell you
what'll cure the huping corf. Do you get some dodmen an'
bile 'em. . . .' "

"Dodmen?"

"Snails," frowned Mrs Heyhoe at this wanton interjec-
tion. " 'An' bile 'em. An' gan the broth to the cheild. That'll
cure t' huping corf.'

"So Jimmy, he got some dodmen, an' biled 'em, an' gan
the broth ter the cheild. But that din't do har a mite a' good.
An' Jimmy, he wor in a rare way, a-wonderin' what to do
arter that, when he run acrost a friend a' hisn, an' tellt him
what was adoin'.

" 'Dodmen? Wor, they ain't no manner a use,' he say to
Jimmy. 'You're barking up the wrong tree altogether. Do
you get hold on a mouse outer a barla' stack. Flare it an'
fry it an' gan the broth to the cheild. Do you see. That'll
cure t' huping corf.'

"Well, Jimmy, there he wor, a muckin' an' messing
about. Lorst two or t'ree days' work.

"Larst he gotter mouse outer a barla' stack.

"So he flared it, an' fried it, an' gan the broth ter the
cheild. But that din't do har a mite a' good.

"There ain't no cure fer t' huping corf," concluded Mrs
Heyhoe triumphantly. "You're certain as there's nothin'
else as you fancy, miss? Then I'll be getting along."

And this time she does, leaving me to sink into a blissful coma mingled with dreams in which incredibly obese mice paced a stately pavane with immensely dignified 'dodmen' on the top of a monster barley stack.

A week dragged by.

In spite of Mrs Heyhoe's unceasing attentions, by Friday I had recovered sufficiently to sit up in bed and cope with the pay book and pay envelopes, which latter Marsham distributed for me. The following afternoon I got up for tea. Two days later I cautiously descended to the office on legs which were just beginning to feel as though they belonged to me again, to deal with some of the arrears of market returns, a pile of accounts and records, and yet another demand from the Ministry of Agriculture and Fisheries for a statement (in duplicate) of all livestock on the farm, and the acreage we proposed to allot to various crops. If it is true that there is a paper shortage, officialdom has yet to hear of it.

Horses, twenty-three. No, twenty-four, because Kitty had her foal last week. A colt, a little beauty, according to Marsham when announcing the new arrival under my window. Though he'd sooner ha' seen a filly. But p'raps the old girl 'ud see her way to putting that right next time.

Cows and heifers in milk, forty-three. In calf, but not in milk, four. Bulls, three. Other cattle, sixty-one. Pigs, none. No sheep, either. (But why are some of them known as two-toothed? And what exactly is a 'gimmer'?)

There is no mention of carrots, with which we intend to sow nearly a hundred acres. The Ministry are evidently much more interested in such crops as Rape (or Cole), Tares, and Kohl Rabi, whatever they may be, and which do not feature on our cropping list. Nor have we any truck with hops, either on statute or hop acres. We get together over grain, however. Wheat, sixty-three acres. Barley, three hundred and one. Oats, forty-four. Green Peas for Market: six acres at the top of Doles, one-and-a-half on Crabb's Castle next Mautby's Hall, all Barn Breck and the Banwell Road Three-Corner, and not quite half the Upper Prisoners, forty-two acres altogether. Mangolds, fourteen. Cabbages, Savoys, and Kale, for fodder: fifteen acres of the last named divided between the Pightles and Leasepit Breck. Cabbages, Savoys, and Green Kale, for human consumption, fifty-one. . . .

"You can alter that to twenty as soon as you like," observed Major Rivers, who had come in a minute or two previously and had been gazing abstractedly at the cropping list. "I'm not at all sure that we oughtn't to knock out Great Horse Close as well. Always follow up a bad market, never a good one. That's one of the golden rules in this game. You watch it. Everyone'll be putting in savoys this year. When the time comes to cut them, I shouldn't be surprised if you can't give them away. We're not likely to get another winter like this last one in a hurry."

"What are we to sow instead?"

"That's the question. You can't do with any more peas. There's no knowing how difficult transport will be by the time they're fit. They're always a bit of a catch crop, in any case. They're not like carrots. They've got to go the minute they're ready or not at all. You might put sugar beet on that strip of Crabb's Castle. There was only about nine or ten acres there for savoys."

"We've got a hundred and twenty-five acres down for

sugar beet already. And our contract is only for a hundred and twenty," I reminded him.

"Heavens above! You don't want to pay any attention to that. You've got to make the best of your land, contract or no contract. A few acres more or less wont matter as long as the Ministry and the Sugar Corporation don't start comparing notes. You've got plenty of beet seed in the barn. There's three parts of a bag left over from last year, I remember Patfield telling me. He wants to plant some of it in the kitchen garden to make sure it's all right, though it's pretty well bound to be. Then he can mix it off half-and-half with some of the new ready for Roger. And barley on the Pound Field. He'll have to use that seed barley off the Little Horse Close if there's enough. If there isn't, I can let you have a coomb or two."

"There's a lot of that Coney Hill seed barley in the corn barn," I ventured rashly.

"So there may be." Major Rivers' smile was tolerant but pitying. "But you don't want to use it on Pound Field. They're both light land. And you should always sow seed from heavy land on light and vice versa. Well, don't over-do it now you're up. There's nothing much for you to worry about at the moment. You've got your savoys out of the way. You can't do anything to Doves Close, Gunspear or the Easter field until we get some more rain. The savoy summary will do any time, and those accounts can wait for payment till next week."

I overslept next morning.

When I woke suddenly and looked at my watch, it was just after eight. Marsham had never missed a morning whilst I was ill. Nothing had happened to him, for I could hear him shouting to Lambert in the yard. Could he have decided rather late in the day that I needed to rest undisturbed? Or had I been sleeping so soundly that I hadn't heard him?

Still, it didn't really matter. There had been no ring from the Major. And it hadn't rained during the night, so there was no need to ring him to ask if there was any change in yesterday's orders. And it was weeks since I had bothered to tune in to the early news. For so long the war had been a sort of stalemate, the two armies skirmishing and snarling at each other like a couple of quarrelsome curs, but neither willing to come to grips. Air-raids were still mostly reconnaissance, with leaflets for missiles instead of incendiaries or H.E.'s. Bombardments were carried on chiefly by words over the ether. People talked of a 'phoney' war, and began to wonder at odd moments if there wasn't a chance of the whole thing petering out.

Soon I should have to start getting up early again. But I had promised myself breakfast in bed for two more mornings. I turned over and slept once more.

"They ha' walked into Norway an' Denmark fust thing this morning, the mucky varmints."

Mrs Heyhoe was standing at the foot of the bed with the breakfast tray.

"*Norway* and *Denmark?* Are you *sure?*"

"Hurry up and come back before Germany pounces on us," Kristin had said, four years ago, as we leaned our bicycles against a fir tree and looked across the bay at Copenhagen.

I had laughed at her.

What should big Germany want with little Denmark, who had never done her any harm, who wished her no ill, and could not possibly have injured her if she had?

Denmark . . . with her friendly little fishmarkets, her clean new factories, her neat farms, a land in which everyone, even the king, rode a bicycle, and you could roam at will over fields and through woods unchecked by notices telling you to keep out . . . to be overrun by a wanton horde of field-grey marauders as ruthless as any of her ancient Vikings.

It had seemed utterly fantastic then. It still did. But it was happening, just the same.

"It's right enow," repeated Mrs Heyhoe stolidly. "I heard it on the seven o'clock. And they give it agen on the news at eight."

I remembered apple blossoms by a village inn, where we had eaten too many cakes, drunk coffee swimming in cream, stayed the night in a spotless room with a heavenly view, and breakfasted next morning, all at a cost of less than five shillings for the two of us. Sunshine on the pine trees and on the dancing sea. That dinner at Aalborg the Svensons had given for me, with eels cooked in eleven different ways. Smorr-brod and good fellowship in many places, with Sigurd, Axel, Sigrid, and Ragnar who confounded all her family with her dark eyes and hair. All the kindness and hospitality I had met with during that halcyon fortnight. The many friends I had made. What was happening to them now?

Impossible to breakfast placidly in bed. Though what could you do when you were up?

Nothing. Nothing at all.

Listen to the news. Listen to Major Rivers saying "Damn the devils. I wish I could get a crack at them. What the dickens are our people about?" Order three hundred gallons of tractor paraffin from the Petroleum Board. Write to Messrs Lingfield and Fenton, Ltd., Importers and Merchants, asking them to send in another five tons of cow meal. Ring up Weldon's to know if they've heard any more about that new sprocket wheel for the W.30. Put down brooms for cowhouse, four, five and six inch nails, and points for the Midtrac plough, on my Meddenham list. Send a postcard to the 'quack' at Westham, who is so much more successful at 'horse doctoring' than any qualified vet, asking him to call in and look at Brisk, who is going lame again. Fill in the dates on the cropping list against the names of the fields that Roger Marsham had drilled during the past ten days. Keep on keeping on, though your mind ached with the futility of trying to make any sort of order in the tragic chaos man seemed hell-bent on creating for his fellow men, because there was nothing else you could do.

"And where did you get to at seven o'clock this morning, Marsham? Couldn't you make me hear?"

Marsham looked sheepish.

"Well, you see, miss, I had a bit of a turn, just as I wor getting out a' bed."

"A bit of a turn?"

"Fust there's a buzzing in me left ear, fares like a hive a' bees. Then I come over all giddy, like. Go down like a ninepin, do I try an' stan' up. So I hed to stay where I wor for a bit, and get Roger to turn out. I'm as right as ninepence now, though."

I eyed him suspiciously. It is impossible for anyone with
Marsham's weather-beaten complexion to look pale. But he
seemed a shade less jaunty than usual, in spite of his as-
sertion.

"What did the doctor say to you?"

"Doctor?" For a minute astonishment and amusement
contended for his expression. Then he was overcome by one
of his chuckling fits. "Wor, I hain't bin nigh the doctor
since one Michaelmas, ten year ago it 'ud be, when I wor
struck with cramp in me right leg. An' I shouldn't ha' gone
then if it han't got so bad that I couldn't put me fut to the
ground."

"He cured you all right, though?"

"That he didn't. All he done wor to ask me a few duzzy
questions, give me a bottle a' med'cin, an' send me off back
home agen. Never said nothin' about the cramp from fust
to last. No more didn't I. I reckoned as that wor his job to
find out. An' if he'd bin half up to it, so he would ha' done,
an' all. It wor my old woman what made out what the
trouble wor in the end. I'd bowt a pair a' britches cheap off
a chap what hed got 'em for hisself up in Norwich, an'
when he got 'em home, they turned out too small for him. I
warn't nothin' nigh as big as him. But it seems as if they
was too tight in the leg for me. I gav' up wearing on 'em an'
I warn't troubled with no more cramp arter that."

This method of consulting a medical man was new to
me, though it is by no means peculiar to Marsham, as I afterwards found out. But obviously something more than tight
breeches were involved in this case: and something ought
to be done about it.

"Have you had any of these attacks before?"

"I ha' hed a few on 'em," admitted Marsham cautiously, "but there, I don't pay no regard ter them. Never know as I ha' hed one onct it's past. I ha' got ter lie quiet time they're on, tha's all. There ain't nothin' else to 'em."

"It wont do any harm to let the doctor have a look at you," I persisted. "I shall be taking the car into Meddenham next week if I don't go this. You'd better come with me and have a word with Doctor Richardson. We can't have you laid up, you know."

"Who's agoin' to be laid up?" demanded Marsham, much incensed. "Whosumever else ha' ter keep at home, it ain't goin' ter be me. Happen I shan't hev another a' these spells do I don't know when. You hain't got no call ter worry about me, miss. I hain't got ter my age without learning a thing or two."

Next day I sought an ally in Mrs Heyhoe.

"One a' these days he'll come over queer when he's on his bike, or summat a' that sort. An' then where'll he be? But there y'are. Billy, he's allus bin as hard to shift as an old dicky, do he han't the mind fer a job. I ha' said my say till I'm black in the face. So's his missus. But fer all the notice he takes, we might ha' saved our breath. I don't know if he'll listen to the Guv'nor."

"Not he!" replied Major Rivers when I tackled him. "Though I'll have a word with him for all that. And you can have another shot at getting him to come into Meddenham with you when you go. But I doubt you won't be able to manage it."

Nor could I. Eventually I gave it up.

The last of the barley was put in, and the first small spears thrust their way up, misting the uneven slopes of Ram's Breck with a pale ethereal green.

Rain fell: not enough and then too much to touch the heavy land. But at last Mark started on Gunspear, the tractor bumping up and down like something on a switchback, with Mark oftener off the seat than on, while behind him the plough tore at the reluctant earth, turning over fragments as large as paving stones.

Small-seeds . . . white suckling, trefoil, and rye grass . . . were drilled over the top of the barley on Sidegate Breck for next year's hay crop.

Major Rivers, looking ten years younger, and as carefree as a schoolboy, organised the detachment of Local Defense Volunteers, which practically every man on the farm joined (or tried to), 'old Tom' being mortally affronted when told that he was too old, as the age-limit was sixty-five, and announcing balefully as he retired unenlisted that do it come ter a scrap, he reckoned as he'd put up as good a show with a pitchfork in his fist as a score a' young 'uns with their tuppenny-ha'penny popguns, an' chance the ducks.

The Upper Prisoners peas were sown, six acres at a time with a week between, to prevent them all ripening at once.

The men took their three-days-holiday-with-pay, to which all farm labourers are now entitled (four in the case of key men or stock-tenders) in batches of half-a-dozen, with the exception of Marsham and Peachey, the former remarking that he'd sooner take the odd half-day here and there when he could nicely manage it, and Peachey mur-

muring that do it was all the same to the Guv'nor, he'd stay
with his horses. "They ont do half for anyone else what
they'll do fer me. An' do they stan' in the stable doin' nothin'
fer four days on end, there'll be no holdin' 'em." ("You tell
that to the ducks on the pit, miss," observed Marsham when
I happened to mention it. "He want the money. Tha's all.")

The sugar-beet land, made ready and then left untouched
for three weeks to let the weeds sprout, was harrowed
again, rolled down, and drilling begun on Swains in an east
wind which blew straight across the sea from Scandinavia
where Denmark had gone under without a struggle, but
Norway was still resisting the field-grey tide.

Names like Trondheim, the Skagerrak, Stavanger and
the Kattegat, were becoming as familiar to English ears as
Southampton and the Solent: and nothing would satisfy
Marsham but that Kitty's foal should be called Narvik. Allied troops were landed, among them Harold Peachey and
Alan Buxton . . . "Though I dunno what Harold thowt
he was a-doin' on, volunteering for a do like that there when
he never could stand the cold. But there, he always wor a
headstrong young bugger," his father said to me.

All too soon the newspapers were talking of 'gallant little
Norway', an ominous sign. Mr Chamberlain announced
that Hitler had missed the bus . . . a sign even more
fraught with ill omen. Within an incredibly short time we
were retreating, withdrawing, and withdrawing again.
The Norwegian Government fled to London, and the
evacuation of British and French troops began at Namsos.
Whilst it was in full swing, the man who had missed the
bus simultaneously invaded Luxembourg, Belgium, and

Holland. Bombs crashed incessantly down on Dutch, French and Belgian towns, reducing them to rubble. Refugees were machine-gunned as they fled. Following the capitulation of the Dutch army after only four days of fighting, came the news of a break-through in the French lines beyond Sedan. The battle of the Bulge was on, with the bulge increasing all the time. Horror piled on horror with such speed that it could no longer horrify. Calamity was a commonplace. Imagination, hopelessly out-distanced by reality, lost the power to torture.

In England the man who had missed the boat at Munich had been replaced by a new Prime Minister. As yet, few bombs fell.

From the office window I looked out on a cool fresh May morning. Pale shafts of sunlight caressed the thousand and one shades of green with which the laggard spring was painting the trees and hedgerows. Beyond the river and the water meadows unsown fields showed dark beside the springing barley or winter wheat.

With a flicker of white tail, a rabbit bounded across the Walnut Meadow and flashed out of sight in the brambles beyond the orchard. Two partridges emerged coyly from behind an apple tree, and began to pick about in the grass. A couple of crows sailed out of the crimson budded walnut branches and winged their way towards Ashacre, followed by the scoldings of a wood-pigeon perched on the ash-tree on the island down by the stream. Somewhere in the distance a cuckoo called.

> "In April, come he will
> In May, he's here to stay."

In the middle of the tennis lawn, a baby rabbit sat up on his hind legs and rubbed his face with two small vigorous front paws.

The cuckoo called again, nearer this time.

Cutting across its song came the shrill wail of Sedgeham's air-raid siren.

Overhead three Spitfires droned in formation, following two thousand feet higher the line taken by the crows.

❧ VII ❧

NO rain fell during the first three weeks of May. The sun strengthened until it shone down on Silford with the heat of late July or August. Soon every afternoon found the young Rivers and one or both their parents swimming in the top pool: and several times I joined them without ill-effects despite Mrs Heyhoe's raven croak of "Bathe in May, be buried in clay, afore the month be out," with which, three times out of four, she greeted my appearance with a towel.

The sugar beet came up, tiny seed leaves of dark emerald which grew brighter and larger every day. Another week and Heyhoe and Peachey, with a horsehoe apiece and Harry Buxton, Cecil Peachey and Oby to lead the horses up and down between the long straight rows set twenty inches apart, began A-hoeing . . . an operation for which A-shaped blades eight inches wide are used to clear the central space between rows of the weeds which have sprung up since the seed was sown, and which must be performed the moment the eye can trace the thin green lines of the legitimate crop across the fields. Before they had finished the ten acres on Crabb's Castle, the North Field was ready to side-hoe: a much more delicate performance since, as its name implies, it takes in the whole space between the rows right up to the sides of the plants, and the least carelessness would result in hoeing up the sugar beet instead of the weeds.

Extraordinary though it may seem, this latter almost never happens. Though, as the plants are larger and their permanent leaves appearing between the seed leaves at this stage, a leader is dispensed with, and the horse is expected to set his course without one. And so they can. "Excepting old Boxer," Peachey told me, when I exclaimed about it. "I dussent use him on this job, do he'd wander all over the place like an old cow. He's a clumsy old varmint if ever there was one. Where's Short he 'ont set a foot out a' turn however long he go."

Ten points of rain fell in tiny showers which barely damped the surface of the ground. The sun was out drying it off again hardly before they were over, while Marsham demanded of a heedless providence what wor the use a' little dags like them there, when we wanted a regular soaker? Plants and weeds grew neck and neck. Mangolds, kale, swedes (turnips), the early peas, were crying out to be A-hoed before the last of the sidehoeing of the sugar beet was finished. Whilst the first sown of the latter were more than ready for 'chopping-out', and every man who could be spared took a hoe and started in the North Field on 'taken work' at fifty shillings an acre for twice.

A wasteful business it seemed to my unaccustomed eye as I watched them hoeing up a score or two of plants for every one they left in the ground. But it was the only way for all that, Will Long, who was in charge of the chopping-out gang, assured me. What with the weather and one thing

and another you couldn't count on a good enough plant do
you put the seed in farther apart. A niceish one this year, it
looked to be, so far. Though it could have done with being
rolled-in tighter. It couldn't be helped. The weather had
been wrong when Roger was drilling it. But it should have
had the Cambridge roll instead of the flat.

"This field want rolling down with a heavy roll. It don't
bind like some on 'em do. You see?" And he let a handful
of the loose soil fall through his fingers. "There's a lot a' the
Manor land same as this here. Diff'rent to the Guv'nor's
over at Ashacre. His hain't got nothing near so much sand
in it. Still, it don't look too bad for North Field, consider-
ing. I ha' seen a long sight worse."

I took another glance.

The plants left behind looked so small and fragile, wilt-
ing under the fierce glare of the sun, that I wondered if any
of them would survive. But next day they had perked up.
Soon they would be ready for 'scoring' or 'singling', when
one of any pair left by mistake would be removed, and yet
another crop of weeds which would have contrived to
spring up between the remaining plants would be hoed out,
since the horsehoe can only work between the rows and
not across them.

"Some farms don't worry about doing 'em a second time.
They just chop out the once and finish with it. But I reckon
it pays," said Will judicially on the Thursday afternoon.
"There ain't all that difference in the price. And I wouldn't
mind laying a bet as it makes a ton an acre difference to the
crop, with nothing nigh so much trouble with thistles in the
barley the year after. We thought of drawing thirty-three

shillings for the chopping out and leaving seventeen for scoring, if that'll suit the Guv'nor. And I'll want to draw for twenty-three acres this week."

And what did I think of the news today?

What did I think of the news?

What did anyone think of the news, today or any of the days just over or to come?

I was inclined to agree with Mrs Rivers, who said that she liked her history best when safely sandwiched between the covers of a history book. To live whilst it is being made, and made at such an unprecedented speed into the bargain, is a nerve-racking process.

In the battle of the Bulge, the bulge had become a gap, a gap forever widening through which tanks poured in an endless stream, axle to axle, sometimes a dozen abreast, obliterating everything in their path, men, women, children, machines. They crashed through Amiens, Arras, Abbeville. They reached Boulogne. The Belgian army surrendered. The epic of Dunkirk began. For the first time since Napoleon planned it more than a hundred and twenty years before, the invasion of Britain seemed not only a possibility but a certainty, with East Anglia scheduled for the initial attack.

Private schools, like individuals, evacuated to Scotland, Wales, Devon or Cornwall (among them West Malton Hall with the three Rivers children), and into the empty houses came troops. Church bells were forbidden to ring again until the Fates sent parachutists or victory. Yet more sandbagged posts sprang up at cross-roads and other strategic points, including the gateway to Ashacre Hall and the

Manor entrance. Bridges were mined ready to blow up and were blocked every night by cars which might have joined in the 'Liberty Day' run to Brighton, decayed wagons, iron hurdles, and barbed wire, with such efficiency that Major Rivers was stopped and challenged twenty-seven times within fifteen miles on his way back from a defence meeting: and John Owles, Arthurton's rival in the bakery and grocery line, who had for years found romance on the far side of Banwell Common undetected by the busiest of bodies, found his private affairs all over the neighbourhood inside a week.

Sea-side resorts were placarded with posters warning the inhabitants to keep bags ready packed with a few necessities and be prepared to leave at an instant's notice. In some coastal areas a curfew was imposed. Inland, steps were taken to render any field or open space above a certain acreage unsafe for landing troop-carriers, and Mark had to take a tractor off to Banwell Close and run a sixteen inch deep furrow across it with the digger plough.

Names were erased or torn from farm carts, lorries, vans, pillar-boxes and village post-offices, and all sign-posts taken down, so that when the enemy landed, they would be unable to find their way about. And from then on, any strayed traveller reduced to asking the way was liable to be suspected of being a Fifth Columnist and denounced to the police.

Mark and Gidney came back cock-a-hoop from Westham station where they had been to fetch a load of carrot bags sent down by various London salesmen who thought the Manor carrot house a safer storing place than their premises in Spitalfields or Boro', with the tale of a gypsy met with on their journey in.

"Stopped us this side a' Wendeswell," related Gidney importantly. "An' wanted ter know the way ter Meddenham. Well, I never see a woman with feet that size afore. Nor a diddykai looking as spruce as she wor at half-arter-nine of a morning. What do you say, Mark?"

"Tha's right."

"We set her on the road all right. Then we nipped into the Wendeswell bobby's as we go by. Reckon they ha' run her in by now."

A tramp who sat down to eat his bread-and-cheese out of an old newspaper on the bank at the bottom of Sandy Loke, shared a similar fate.

I very nearly joined them myself.

I had been into Sedgeham to fetch a new gasket for the 'crawler', and on the way back mistook a turning. Bearing left a little farther on, hoping to get back on to my original road, I came instead to a four-cross-ways, and plunged for the wrong track again. Soon I was hopelessly lost. I stopped a small rabbit-faced man on a bicycle.

"Can you tell me the way to Silford, please?"

There was a lengthy pause, during which Amy and I were treated to a long searching stare from two inquisitive eyes above a twitching nose.

"The way to Silford?"

"Yes, please. I want to get to the Manor farm."

"Well, I know the way to Silford."

Another pause.

"But I dunno as I owt ter tell you."

"But I've got to get back to Silford Manor; and I've lost my way!"

"I don't know nothing about that."

It occurred to me that he didn't know the way to Silford, either, but wouldn't admit it. However, I couldn't be very far from Sedgeham. And if I could once get back to it, I could soon find the right road again. Perhaps he would be kind enough to direct me to Sedgeham instead?

"Well, I know the way to Sedgeham."

We shared a two-minute silence, during which my fingers itched to grasp a spanner.

"But I don't know nothing about you. An' where's the good a' the Guv'nment taking down all the signposts, an' then me a-telling everyone where ter go what asks the way?"

I'm afraid I didn't know the answer to that one. But Mark was waiting for the gasket. And I ought to have been back with it by now. I pressed the self-starter.

"I'll tell yer what!" My tormentor's voice suddenly squeaked above the throb of the engine, and I switched off again. "If you tek the furst tu'n left an' then tu'n right in about an hundred yards, you'll come to a house on the left-hand-side a' the road. Tha's the policeman's. Do you go an' ask him the way. An' if he think as you owt ter know, he'll tell yer."

As Amy moved off he was obviously memorising her number.

Bombs were scattered up and down the country, though most fell in East Anglia: and human nature being what it is, the first casualty was the Ministry of Information-spon-

sored Sixth and Silent column. After all, wasn't freedom
of speech one of the things we were fighting for? People
wanted to know. And plunged with joyous abandon into
the game of 'Where did that one go?'

For the time being, while houses rocked, windows rattled, now and again bits of plaster fell from ceilings, or a
tile dropped off a roof, and every morning each of us was
prepared to swear that the enemy planes had spent the entire night circling round and round over his or her individual head, Silford scored nothing nearer than the thirteen small H.E.'s which fell in a hayfield just beyond Banwell Common without exploding, two of them being personally discovered by Heyhoe by the simple expedient of
poking a pitchfork down every likely looking hole. But we
all joined in blithely just the same, eagerly comparing notes
as to time, number, and rocking power of anything going
'bump in the night', and demanding of all and sundry next
morning "Where exactly did they drop them this time?" A
question to which an accurate answer was practically always forthcoming before mid-day. Lesser wights might be
satisfied with the newspapers and the B.B.C.'s vague refer-
ences to 'an East Anglian town', or Random, that elastic
and much attacked village. Thanks to the Milk Marketing
Board lorry driver, the tradesmen's vans, the odd engine
driver or railway porter contacted by Mark and Gidney, or
George Drayton when he collected his papers at Dennington Station, and the L.D.V.'s who manned the bridge by Silford Pool and the shepherd's hut perched on the top of the
hill near Gypsies Pit each night, we k-n-e-w.

"You're sure you're still all right on your own?" Mrs

Rivers asked me, soon after the Banwell hayfield episode.

"Good gracious, yes! Nothing very dreadful's happened yet."

"You wouldn't rather sleep downstairs? We've taken to doing it lately. And Nicholas' study could soon be fitted up for you."

Though I don't in the least mind being alone in the house upstairs, I have sundry qualms about sleeping on the ground floor with open windows. Besides, I knew the study. I much preferred my flat.

"Well, please yourself. But I think you ought to have something done to those windows of yours if you stay upstairs. Charles brought that Home Office research man who's come down to look at the Banwell bombs in to tea yesterday, and was asking him a few questions. He seemed to think you are almost as safe upstairs as down, provided you have stout wooden shutters on the windows and keep them closed if there's any fun going on. You merely have rather less time to get out if a small bomb comes through the roof. If it's a large one, presumably there isn't enough of you left to worry anyhow. There's plenty of wood in the carpenter's shop, isn't there?"

"A fairish amount. Thompson's sent in our June ration two days ago."

"Then I should get Peacock on to it, right away."

There are two windows in my bedroom, so I could always keep open at night the shutters not directly behind my bed. But even so it seemed a stuffy idea in the lovely weather. The chances of a bomb on the Manor must surely be something like a million to one against, especially now that the long light evenings and 'summer time' had practically put an end to complaints about lights. Even Gidney, who is an A.R.P. warden, had ceased to beg me to speak to Lambert again about his blackout. And didn't glass always fall outwards?

I put off speaking to Peacock for a day or two.

Towards the end of the week I woke to find the house walking off towards the Walnut Meadow. Then it turned round and marched back again. At least, that was what it felt like. In about ten seconds I was out of bed and had my clothes on.

Lately I had amused myself by estimating the distance between the Manor and the latest bomb to fall, and had managed to place most of them within half a mile within a radius of up to twenty miles. This one must be just behind the corn barn. The 'wah-wah-wah' of the plane which had dropped it sounded immediately overhead. I waited for the next crash.

The house was still shaking.

No, it wasn't. It was me. The sort of thing one never dreamed could really happen, except in novels. Oh, for the sound of a human voice! I snatched up the telephone to dial Ashacre Hall. Then put it down again and snatched up Master Pepys instead, who purred unconcernedly and so loudly that for a moment he drowned the noise of the plane engine.

Thump. Crash. Thump. Crash. Thud. But farther away.

Only the house shook this time. Samuel Pepys still purred. Together we leaned out of the window and listened to the Croft and Bailey families exchanging pleasantries on the strip of meadow between their cottages and the cow-houses.

"That fust one was a master rum 'un, warn't it, miss?" called Croft, who had heard my window open. "Out by Gilderstone by the sound of it."

Gilderstone is a good five miles from the Manor. I couldn't believe it. But so it was. There must be some peculiarity in the lie of the land, a seam connecting part of Gilderstone with the Manor on the lines of the whispering stones found in France, for a small bomb there will always shake the house twice as much as a heavy one a good deal nearer home. This particular one weighed five hundred pounds, and in the morning two or three householders at Wendeswell reported odd broken panes with the glass fallen inwards, though the rest of their windows were intact. So, it turned out, were most of Gilderstone's, including those in the fifteenth century church on Gilderstone Hill, barely half a mile from the explosion.

On my way back from Sedgeham a day or two later, having taken a couple of calves into market with Amy and the trailer, I made a slight detour to look at the crater which jungle telephone had already correctly reported as being in the middle of a sugar beet field on Gilderstone Church Farm, and big enough to hold a couple of haystacks.

Chunks of grey clay subsoil, chalk, and sand had been flung in all directions across the thirty-acre field, and even over the hedges into fields adjoining. Inside, the crater was terraced with minor peaks and valleys. Two huge silver-grey metal discs, the size of millstones, leaned drunkenly against the torn earth near the centre.

The effect was neither frightening nor menacing. Nothing spoke of the hideous death which would have come to a score of innocent people had the bomb's release been delayed a single second. Instead the whole thing seemed uncannily detached and remote, as though it were in a different world. I might have been standing beside a burnt-out volcano on the moon. I scurried back to Amy.

That afternoon Peacock came in to fit and hang the shutters he had been making for the flat. I was having tea, but I knew better than to offer him a cup this time. As he had brought me a bottle of his home-made wine, I fetched two glasses instead.

"Bullaces. Tha's what tha's made of," he told me as we sipped. "This here's goin' ter be a rare year for ston' fruit, do you see. That fence along the top a' Booters, next to Easter Field, 'ull be creepin' with bullaces. I doubt, though, they 'ont let me have 'ny extry sugar fer 'em, same as they're a-doin' fer jam. Do me a sight more good. Not but what the finest way a' makin' off with bullies is with gin, do you can come on the gin. Hain't you never tried it, miss?"

"No. Is it good?"

"A master rum un," said Peacock reverently. "I larnt the way a' doin' it time I was doin' a bit a' shepherding in the Cotswold country onct. You put bullies into an earthenware crock with the gin, then cork 'em up an' stan' 'em down by the hearth an' give 'em a rock every night fer a year an' a day. Then you bottle it off. That'll warm the cockles a' yer heart an' no mistake. The business is gettin' a holdt a' the gin. Time was when a sailor chap what I come acrost useter get it fer me fer five bob a bottle. Ar, an' it wor gin, an' all. But I ha' lorst track a' him, so I hain't made any fer dickey years now. Well, this 'ont do." He put down his glass and picking up his screwdriver, vanished into the bedroom.

The first half of the shutter fitted perfectly. The second was somewhat wide.

"Now how the go-to-flyin' did I come ter do that?" marvelled Peacock. "I could ha' sworn they wor buth alike. Measured 'em two or t'ree times an' all. Ain't that a masterpiece?"

Do I could just steady it fer a minute, though, he'd soon nip a bit off yon side.

"Tha's the ticket." He stood back to admire his handiwork. "Inch an' a half board. That owt to keep the glass off you, do they let fly another a' them solers same as that one at Gilderstone."

"I had a look at the crater on my way back from Sedgeham this morning."

"Me an' Isbell, we cycled over and took a look yesterday evenin'. Not muchers where aiming's consarned, are they? Slapping it down in the middle a' a field when they might ha' got the village an' the chutch. I reckon as old Harry must ha' bin prowling round the chutchyard an' caught an' hulled it off the premises, same as he do ter folks what sticks pins in the gate-pust."

"Who do what?"

"So I've heerd tell, time an' agen, though I hain't never tried it meself. Do you stick a pin in the gate-pust a' the big entrance gate on the road, not the littul totty one round ter the side, an' then walk three times roun' the chutch outside, bendin' down each time you pass it ter look inter a square hole made by a brick or two what's missing down near the ground in the far side a' the chancel wall, you'll never be able ter get outer the chutchyard the same way you come in. You'll be hulled over the wall an' downhill inter the medder afore you know where you are. Do you try it one a' these days an' see if that ain't right."

So I shall, though I haven't yet.

"They say as it all started along a' an old woman what used ter live near the chutch, an' kep' toads in a teapot up the chimney. That 'ud be 'bout seventy or eighty year ago. She put a curse on the blacksmith, then dared him to go an' do it. An' sure enough, over he go. An' the same thing happened to some more on 'em, what had a go arter that. Then she hed ter go an' put a curse on his children an' all."

"Whatever happened to them?"

"Come home from school the very next day with their heads alive with nits," said Peacock solemnly.

"That's been known to happen without anyone putting a curse," I suggested.

"I don't know about that. Tha's what happened to 'em anyway. She used ter be able to fetch people, too. Reckoned she could fetch 'em any distance, without sayin' a word or nothin', just by sitting in her front room. But the blacksmith, he stopped it in the finish. He pricked her with a needle an' fetched blood. She couldn't do it no more arter that."

The second pair of shutters were a perfect fit, but the hinges had unaccountably been left behind in the carpenter's shop, and a special journey had to be undertaken to retrieve them. The sitting-room shutters somehow managed to be too big altogether.

"We'll soon put that right," announced Peacock, getting his plane out of his tool bag. Soon the floor was covered with shavings.

"Mind you, I don't hold with meddling with them kind a' things meself," he said abruptly, still planing busily. "Time was when I din't know no better, an' thowt as I'd see what it wor all about. There was a chap what used to live on Banwell Common time I was a lad, what could stop horses. There they'd stan' an' couldn't move till he give the word. An' I ha' see a stable wall fly afire when he scratched it with a two-tine fork. Then he'd scratch it again an' put the flame out. All along a' his havin' the bones a' a walking toad in his pocket. So in the finish nothing 'ud suit me but I must ha' a go.

"You can't make a mistake over the toad. It's darker than ordinary, with an arrow on its back. Though I was a master long time afore I ketched hold a' one. I might ha' know from that as it wouldn't do me no good.

"Howsumever, I pegged it down atop a' an antheap, an' give it a week. An' then I come back fer the bones, which was picked clean as a whistle. The next thing was ter tek 'em along to a running stream an' drop 'em in by a bridge at midnight. The ones you want are them as float agin the stream. An' you hev' to look sharp an' collar hold on 'em afore they float away. Then you mustn't never leave 'em, always keep 'em on you, do they 'ont work.

"I knew as the bridge 'ud start roaring like thunder the minute I hulled the bones in. But I hadn't never reckoned on it being as bad as it wor. I wor only a lad at the time, as I said afore. An' the long an' short of it wor, I cut an' run as though old Harry hisself wor arter me, when the row started. An' I never went back no more. Now, me beauty, le' see how you're agoin' on."

This last sentence was addressed to the shutter, which declined to return the compliment, however, and now proved to be a fraction too short, to Peacock's discomfiture.

"It'd be a funny bit a' glass that'll get past it, all the same," he asserted doggedly, at last. "Tisn't as though you're agoin' to sit atop on it, neither, is it, miss?"

It certainly isn't.

"Then I doubt as she'll do. I ha' made an hour, if you'll put it down in the book. An' now I'll be getting along. My second sowing a 'taters is just about ready to mould up. Ernie's fust lot aren't fit for it yet, I see."

The evacuation of Dunkirk ended. Officially it was a moral victory, a masterly withdrawal, a miracle. No doubt it was. Though officialdom was reticent to a degree about the details, the stories which reached Silford through relatives and friends who had manned Wells cockle-boats, Yarmouth fishing smacks, holiday craft from the Broads, anything and everything which could float, would have filled a dozen volumes with tales of courage, endurance, the incalculable workings of the thing called luck, even laughter, and then touched no more than the fringe of it.

Telegrams came to Silford and Ashacre, some from the rescued telling of their safety, others from the War Office, regretting to announce.

Mrs Croft had a letter from Mrs Dack, who on one pretext and another, had stayed on with Elsie in Leeds. Dack had been safely evacuated, but on the way from quay to camp, the lorry in which he was riding, overturned. His widow was making her home with Elsie for good now. Somehow she couldn't fancy coming back to Silford to pack up her bits of things. Perhaps Mrs Croft would be so kind as to do it for her. Geoffrey Blake was shot down as he bailed out of his falling plane over Calais. Of Captain Irstead there was no word.

The two Buxton brothers had started to swim out to the Lancastria together. Alan was always the better swimmer, 'old Tom' told me with wistful pride. Bertie had told him to get along: he'd see him later. When the ship was hit, Bertie was some four hundred yards away. He had time to swim back before the enemy set fire to the oil-covered water full of struggling men. Then a motor-boat picked him up.

Jeremy Peacock, looking incredibly tired, but wearing an irrepressible grin just like his uncle's, appeared in the yard one afternoon, having given himself leave from the camp 'somewhere in England' where he was supposed to be re-cuperating from his adventures, and hitch-hiked to Silford in four-and-a-half hours. He had spent three days and nights on Dunkirk beach. The seven previous to that his unit had been constantly on the move, setting up their guns in each new gap, firing, moving on again. The only thing that seemed really to worry him was the loss of his ciga-rettes.

"Nigh on two thousand on 'em which I'd boned, seeing as they wor lying around, an' I reckoned as I might just as well have 'em as Jerry. Kep' 'em safe until right the last.

Tied 'em atop a' me head afore I started wading for home, so's they'd keep dry. An' then, blow me if they didn't drop off just as I wor clawing up the side a' the boat what brought me back," he said to me indignantly when I came across him loitering in the hayfield next day, watching Peel sharpening the cutter knives.

Normally the hay is not cut until after mid-June. Many farmers refuse to touch it until the longest day is over, a procedure of which Marsham is all in favour, though he allows that there are exceptions. This year was one of them, and on the Manor farm haying was nearly a fortnight early, though the crop was what Marsham described as a 'funny' one, being a record for quality and quantity, in spite of the season.

Up on Palegate Mark with one Fordson towing a three-wheeled iron-frame Barnlett grass-cutter, and Vincent with the other trailing a McCormick reaper, were laying low swath after swath in ever-decreasing circuits, while Myhill lurked about retrieving baby partridges which ran helplessly up and down cheeping as the sheltering grasses fell. Soon they were moving into the Well Piece.

For a week the hay lay drying in the fickle sun, turned over occasionally by Peachey and Roger Marsham with the horse drawn 'topplers'. Then the men, working late into the evening . . . for the heavy morning dews and the extra hour of 'summer time' made it impossible to start early . . . cocked it into heaps to 'make' before stacking.

"They wor sayin' in the 'Hart' las' night as France'll pack up an' all within the next four-and-twenty hours," volunteered Peel, the morning they started in the Well Piece, as I stopped to watch him at work.

"FRANCE?"

Despite the evacuation of nearly three hundred and thirty five thousand men from Dunkirk, there were numberless British soldiers fighting on the other side of the Channel, and reinforcements were being rapidly landed. Though Italy had declared war on the Allies, so far she had done no more than raid Malta, to which the R.A.F. had replied by bombing air-bases in Libya and military objectives in Genoa and Turin. And while the German advance had come perilously near to Paris, its citizens had declared themselves ready to defend their capital street by street, and already the French armies were counter-attacking with some success.

"It can't be true. There was nothing about it on the news. Besides, she's pledged not to make a separate peace."

"Do tha's what I heard. An' Catton, he'll tell you the same. His brother-in-law what works at Wroxham and went over to Dunkirk with some more on 'em, he said then as he wouldn't give 'em more'n another fortnight. Now all the talk is as they'll ha' packed up by tomorrow."

Two days later it was officially known that Marshal Pétain had asked for an armistice. Mr. Churchill's offer of a Franco-British Union had been refused. Before the week was out the French Government had accepted terms amounting almost to complete surrender. Another and greater evacuation problem faced the British Army. Only a narrow strip of water prevented the German tanks from sweeping straight on to England. The British public expected invasion within the next few days: and Mrs Rivers, already frantic with anxiety over the fate of her brother, decided a dozen times a day to fetch the children back from

Cornwall before she eventually left them where they were, and organised the local group of the 'stay-put' talks which were being given everywhere to prevent a repetition here of the hideous French and Belgian refugee débâcle.

The Blitzkrieg on Britain began. At the Manor, the work of the farm went on.

Battles might rage, countries and bombs fall, personal anxieties cut nearer the bone than shrapnel splinters, but the sugar beet and mangolds must be singled, wheat and barley cleared of docks and thistles (the women were doing it this year under Patfield), the baby carrot plants inspected every day lest they should be attacked by 'flies', when Kirk must be immediately despatched to broadcast soot over them, the kale and savoys put in, the pigeons chased off the peas, and the hay carted and stacked . . . though not in the middle of the fields, as recently suggested by the Ministry of Agriculture with a view to discouraging troop carrier landings.

"Do how do they suppose as we're agoin' ter cart it away agen without muckin' up the next crop?" demanded Marsham scornfully. "Look pretty, we should an' all, fetching a tumbril slap acrost growing corn or roots every time we want a bargain a' hay fer the animals, for the next twelve-month. I lay you a bet they never thowt a' that."

While there is no actual partnership between the brothers-in-law, and both the Hall and the Manor are run as entirely separate units, it has always been an understood thing that whenever necessary, one or the other can borrow either men or implements or both. When the last blades of hay had fallen on the Well Piece, Mark had taken his tractor and the Barnlett and joined the Major's haycutters. Now he went off again and brought back the Mugleston Sweep, a species of gigantic rake·which Major Rivers had bought at the last Agricultural Show. Next morning he took it into Palegate.

It was a fascinating performance to watch.

Propelled by the tractor, the Mugleston gently charged a haycock, scooped it up as easily as though it were no bigger than a mole hill, and then passed on to the next. When it had collected four or five, Mark swung it round and sallied back over the field to where an outsize stack was rising . . . thirty-six yards by six, from thirty-five acres.

I stood wondering how it would dispose of its burden when it reached the elevators.

Would it toss the hay up and off? Turn over and drop it upside-down? Tip it off sideways?

It did none of these things.

Mark simply backed the tractor, and the Mugleston slid out from under the haycocks, leaving them sitting there intact for the waiting stacking companies to pitch on to the elevators. Then he sailed off down the field again to fetch the next batch. The whole operation had taken only a few minutes, and had the finished perfection of absolute simplicity. Incidentally, with one man and a tractor, the Mugleston can accomplish more work than eight men, six horses and two wagons, and keep two elevators and stacking companies fully occupied.

I stopped a moment beside the second elevator to watch the hay being carried up on to the stack. Most of the pitchers had been up all night, for another invasion scare was on, and the L.D.V.'s had been standing-to since 4 A.M.

"Don't you think as it's about time you learnt to handle

a gun, miss?" enquired Will Long, as he rested on his fork for a moment. "You never know when it might come in handy these days. There's always some of us practisin' in the pithole at the bottom a' the heath between seven an' nine of an evenin'. Why don't you come an' hav' a go some night?"

I had never handled a gun in my life before. I have qualms even over the killing of a wasp. But eight o'clock that evening found me lying on my stomach fifty yards from the sand-bagged end of the gravel pit, with a .22 rifle against my shoulder. I still have my first target, a square of cardboard in the centre of which is a black circle double-ringed with white. I was the third or fourth to use it, and the three holes inside the outer ring, another half in the black, and the chipped top right hand corner, all carefully marked with red pencil, record my first five shots. On four subsequent visits I scored a bull at twenty-five yards, a respectable number of hits in the 'white', several gorse bushes well off the mark, and three sandbags. But I managed to tie with Will Long in a match of ten shots apiece, to his great disgust and the delight of the pack of small boys who were always hanging around.

"There's that .410 of Nicholas' in the gun-cupboard in the office," Major Rivers reminded me the day after the Germans landed in Jersey. "It wouldn't be a bad idea for you to keep it up in your 'flat'. Ask me for some ammunition for it when you come to tea this afternoon."

Instead of visiting the gravel pit that evening, I hung up three old envelopes on the box-bush and the laburnum tree at the side of the tennis lawn, and took the .410 upstairs.

"Remember to challenge three times before you fire, or

it's murder." And "Don't make a target of yourself at the window, or they'll have you before you get them," had been the sum-total of the Major's advice when he had shown me how to open and close the breech a few hours earlier.

No one, not even Will Long, who had constituted himself my mentor in all matters ballistic, had ever mentioned to me the importance of looking down the barrel before firing, to make sure that it was clear. The last person to use the Manor .410 had put it back in the gun-cupboard without cleaning it. Crouched down below the windowsill, I took aim and pulled the trigger. The gun 'kicked'. The next moment I was spitting out fragments of front tooth, and racing into the bathroom to press cold water bandages on a rapidly swelling upper lip. Down on the box-bush an 'enemy parachutist' flapped untouched in the breeze.

If and when invasion comes, no one can be certain exactly what he or she will do.

Meanwhile the .410 is back in the gun-cupboard in the office. Occasionally I take it out pigeon-scaring. But on the whole I prefer to leave this chore to the clock gun, which Myhill stands out in the field and sets to fire automatically a shot every hour. It is much more efficient. And it has no teeth to worry about.

❧ VIII ❧

THE flies kept off the carrots. On Prisoner's Close
and the Pightles, however, rabbits strayed in from
the glebe fields wedged between the Hart Piece, and
Coney Hill, which, owing to the way it was farmed (or
neglected, rather) by the smallholder who hired it, had for
years been a running sore from which thistles, docks, and
every variety of weed as well as vermin, drained into the
healthy body of the Manor land. Where was the use in
gassing up an' down our side a' all the fences time an' agen,
when Lopham, he never did nothing about his, a disgruntled Patfield asked again and again. Captain Irstead had
lodged repeated complaints in the past. Major Rivers added
to their number in the present. Marsham fumed. Myhill
paid sundry visits with 'Rags' and a gun. Kirk scattered
bags of soot over the devastated areas. For the time being
there was no more to be done.

On Leasepit Breck the 'flies', in this case hordes of small
insects resembling black beetles, with yellow stripes down
their backs, ate off six acres of kale before the plants were
past the seed-leaf stage. The savoys at the top of Banwell
Close succumbed to the drought. Both had to be rotored up,
harrowed, and rolled by Vincent and a tractor, more superphosphate sown by Kirk, and then redrilled by Roger
Marsham.

The Petroleum Board rang up from Sedgeham to say that a military order had come through forbidding the storing of petrol in private pumps, and announced that they proposed to empty out anything left in the Manor's at once. At nine o'clock that evening a lorry departed with the seventy-eight gallons of pre-war petrol which had been Croft's pride and joy. Next morning Cecil Peachey, in obedience to another official pronouncement, spent the best part of half a day filling up the pump with six hundred gallons of water. After which he spent a day and a half helping Kettering to make several dozen branch brooms and set them up in bunches at convenient intervals around the various cornfields, in case of incendiary bombs. "Though I doubt as we 'ont get much trouble, do they do drop 'em," the latter remarked when I came across him in Rams Breck. "Wheat 'ont burn at all, as a rule. An' barley not as easy as you'd think for. An' then not afore noon, once the dew's off. I made out all about that when I used to work out yon side a' Meddenham, on a farm what had the railway running clean through it. Sparks was always setting light to the fields what run alongside. But they never did no harm worth mentioning except once. An' that was barley, at three o'clock in the arternoon."

Horsehoes and 'scratchers' manned by Peachey, Catton, Heyhoe, Hill and Ted Long went up and down the rows of growing carrots, sugar beet, kale, swedes, and mangolds, over and over again. The first named were also weeded with 'crawler' hoes, short handled implements wielded by hand with bodies bent double to remove thistles, embryo docks, grass, birdseye, and seedling 'fat hen', a weed which in an incredibly short time turns into a bushy plant often over a couple of feet high, though it at no time looks like an over-fed fowl as far as I can see, nor would any of Marsham's score or so Rhode Island Reds touch it when I dropped some over into their run.

Bushell, with the 'sail-cutter', trimmed all the pastures and some of the upstream meadows. The Silford Hill cattle, having eaten everything off those downstream, were moved to Kidholms (where I could count them from the flat window) staging a twenty-minute rodeo in the Ashacre road on the way. Agricultural wages took a flying leap up of ten shillings a week, and Mrs Cutbush launched a new War Savings campaign with house to house visits. I stamped up everyone's Unemployment and Health Insurance cards, posting the former in a bunch to the Sedgeham Labour Exchange, and handing out the latter to each individual employee on pay day. The peas ripened.

"Kate and I are off up to London to badger a chap I know in the War Office for news of Nicholas. Then we're going on down to Cornwall to take a look at the kids," Major Rivers told me just before the first peas were ready to pick. "The hay's out of the way. Harvest won't be here just yet, though you'll find it will be early this year, and we shall be cutting oats well before the last of this month. Meanwhile you've got nothing much to worry about except the peas. And you should be able to manage them without me between you."

"I've never done anything but shell, cook and eat them before," I murmured dubiously.

"No, but you've had plenty of experience with the carrots. As far as the markets are concerned, you work them exactly the same. Look up in the book to see what each man had last year. And then send him more or less according to the price he makes of them. Barrett's will cart them up. The only thing is that you'll have to watch them like a hawk. This hot weather is bringing them all on together. And if they're ready it's no good leaving them a minute. You know yourself the sort of peas you'd buy, and the kind you wouldn't take on as a gift. So do Patfield and Marsham. If they come on too fast for the women to keep pace with, you'll have to set on all the men you can spare as well."

Three days later I saw the Rivers on to the afternoon express to London from the platform of the main station of a not-far-distant 'East Anglian town'. The petrol rationing and the absence of signposts had made them decide against making the journey by road. The rather recent enemy attention to the East Anglian town in question had caused Major Rivers hastily to rescind his original intention of driving up in the Vauxhall and garaging it there whilst they were away. So in the end Amy and I brought them in. As the train drew out, the sirens were sounding again, but nobody paid the slightest attention to them. "The only times anything happens, there's never a warning on," the collector grinned at me as I handed over my platform ticket. Outside the station the streets were so full that it was several hundred yards before Amy could be allowed into top gear.

When I was nearly home again, I turned right opposite the turn to Silford Hill and went round by the Banwell Three Corner Field, where pea picking had started that morning.

I had arranged with Barrett's to take three hundred and twenty bags that evening, and already nearly three hundred of them were neatly stacked in heaps according to the consignee. Ranged over a strip of the field about forty women were busily tearing up the plants and stripping off the tight green pods, dropping the latter into the big wicker pea skeps, and letting the former fall back onto the ground where the children who had accompanied them built them into 'castles' and played happily in the sun.

As each skep was filled, it was brought across to Patfield, who emptied it on to the potato-riddler, and then pushed the pods down into the waiting sack fastened at the end of it. If there were more peas than the sack would hold, the remainder went back into the skep towards the next bag. If the steelyards which depended from the large iron tripod standing opposite the riddler failed to register forty pounds when Patfield hung the bag onto the hook, then the picker had hurriedly to gather or borrow enough pods to make up the correct weight. Not until then would Patfield part with one of the lettered and numbered tickets which were handed over for each full bag. This must be produced again on Friday when it would be redeemed for one-and-three-pence (last year the price had been a shilling) and during the weekend I should have the pleasure of sorting them all out again into sets numbered one to a hundred, each with their initial letter, ready to use over again the following week.

"When did you last have a look at the Upper Prisoners, miss?" enquired Patfield, as I handed him over the consignment notes for Barrett's drivers.

"On Friday afternoon, with Major Rivers. Why? What about them? The first lot didn't go in until a fortnight after these. They'll be all right until next week, won't they?"

"Tha's what I thowt a' Friday. But I come round that way this morning and they're come on a rum un over the weekend. They'll want pulling, some on 'em, by We'n'sday, Thursday at the latest. And we shan't ha' cleared this afore Friday night, if we do then."

"The Major thought something like that might happen. If it did, he said we were to start the men off. I'll tell Marsham."

When Major Rivers had mentioned it, the whole thing had seemed absurdly simple. A mere matter of issuing an order. In conference with Marsham, however, it suddenly dawned upon me that the situation was delicate in the extreme, needing the wisdom of a Solomon for its solution.

"How much are the men apicking 'em for?" asked Marsham, when I had said my little piece.

"How much? One-and-threepence a bag, of course. What else should they get?"

"They 'ont be able to arn their wages at that price, not now they're gone up same as they hev."

"They'd better work on day's wages in the ordinary way, then, I suppose," I murmured, after a pause.

"That 'ont suit neither, do you see. You 'ont get half the number a' bags pulled, do there ain't something extry hanging to it. They 'ont be able to pick as fast as the women do, neither, not being used to it. I reckon as it'd want to be something between one-and-a-tanner an' one-an'-nine a bag to make it wuth while for 'em to put their backs into it."

"And what do you think the women would do if they heard that the men were getting up to sixpence a bag more than they were, for exactly the same work? They'd all strike. And I shouldn't blame them, what's more."

"Well, there y'are," said Marsham helpfully.

In the end the best I could do was to pay the men by day's work, and add a bonus of threepence for every bag picked (subject to the Major's ratification), trusting that none of the women pickers would prove mathematician enough to work it out and spread the glad news. While to discourage odious comparisons as to the respective picking capabilities of the sexes, Patfield and the women stayed on in the Three Corner with orders to move up to Doles if they finished first: and Will Long took charge of the men, and started on the lower end of the Upper Prisoners on Wednesday morning. At the end of the day their average was something like two bags per picker below the women's. Next morning, to complicate matters already complicated enough by the fact that it was no good picking after Friday, and Heaven alone knew what the peas would be like after the weekend, it began to rain. When I arrived at the Upper Prisoners shortly after ten o'clock, having bumped and bounced Amy down the track which runs at the bottom of Coney Hill, Pound Close, and the Pightles, with a trailer full of sacks, not a man was to be seen picking.

"Didn't any of them turn up this morning?" I demanded of Marsham, who had ridden down with me.

"A course they did. An' all on 'em pea pullin' when I come this way afore brekfust. Now where ha' they got to? Do they're going to make off same as this here, you'll ha' to cut 'em off. It ain't rainin' nothing to speak of, not now. Roger, he's bin at work all the time. So's the rest on 'em what ain't supposed to be up here."

We crested the hill and walked down towards the flint-walled yard and premises which stand on the edge of the Upper Prisoners at the top of the lane running up from the Horse Meadow, and are known as the 'New Buildings' though they are older than the memory of anyone on the farm. The door of the turnip house stood half-open. Inside the delinquents were absorbed in shove-ha-penny.

"So this is where you ha' got to, together!" said Marsham austerely.

Outside the rain suddenly quickened from a drizzle to a downpour.

"Can't do much in this," suggested Will Long placably. "We keep on dodging out the minute it eases off. An' I'm keeping a count a' the times."

"So I shud hope," returned Marsham, but less severely this time.

A silence fell, broken only by the swish of the rain outside.

"I shall have to be getting on," I remarked finally. "What about you, Marsham? Do you want a ride back?"

"I may as well stay an' hev a look at them yearlings on Three Bridges now I'm here," said Marsham thoughtfully.

As I rounded the yard I heard the faint but unmistakable chink of a coin rolling over the brick floor of the turnip house.

Friday was fine, and Barrett's carted off four hundred and fifty bags of peas to the London markets. The week-

end was showery. On Monday it rained again, but mercifully the weather had turned cooler. The 'Evergreens' on
Barn Breck, which Patfield had gloomily prophesied would
be bursting out of their pods by mid-week, were to take another ten days to ripen. The 'Thomas Laxton's' on the Upper Prisoners no longer tried to race the earlier sown 'Fore-
most' to maturity, but settled down to reach it at a steady
jog-trot. By Wednesday I was able to send the men back to
carrot hoeing, and let the women take over alone.

The Rivers came back, full of Cornwall, not quite so
talkative about London, though they had been successful in
obtaining news of a sort. It was almost certain that Captain
Irstead was a prisoner of war, Major Rivers told me as we
walked round the farm next day. Jameson was still ferret-
ing round, and would let them know as soon as he could get
hold of some more details. Meanwhile, it might have been
a good deal worse, though naturally Kate was very much
upset.

We were standing on the crest of Sidegate Breck, where
we had come to see how the small-seeds meant for next
year's hay crop, were faring in the barley.

Below, the field and meadows ran gently down to the
river's edge. Beyond it more fields and meadows, with here
and there a cottage, sloped upwards to where woods, indigo
touched with purple, leaned against the sky line. Down to
the left cows browsed contentedly on the lush grasses beside
the willow belt. To the right, 'Blue Peter' stood by the gate
leading into the Three Bridge meadow, flicking off impor-
tunate flies with his long black tail. In the empty sky overhead there sounded the queer muted note of that most accomplished of nature's ventriloquists, the nesting swan.

So must the Manor's owner have stood, not once, but many times, looking down perhaps from this very spot, drinking in the valley's deep inarticulate peace. Now, imprisoned by the enemy as well as his temperament, he was caged and confined as hopelessly as the big red bull up in the yard in his narrow brick stall beside the cowhouse, without the latter's occasional solace of breaking down a wall or door, or waking the echoes with a mighty bellow. I couldn't help thinking what a pity it was that his sister could not change places with him. No prison camp, whatever its conditions, could have held her for long.

"No use worrying," said the Major abruptly. "Good job we've had these rains, anyway. We could have written this off as far as next year's hay is concerned, if the drought had gone on much longer."

The drought had brought on all the soft fruit early, too: and the Ministry of Food was urging patriotic rural housewives to organise preserving centres to avoid waste and help the nation's food supply. Mrs Rivers called a special meeting of the Silford and Ashacre Women's Institute, at which it was decided to join in the scheme; a committee was elected, an allotment of eight cwt. of sugar was asked for (and obtained) and members gave up sundry evenings to make jam which was to be on sale at cost price or thereabouts. They also tried their hands at canning, an official demonstrator coming down one afternoon to show them how it should be done. But this activity was limited owing to the difficulty of borrowing a canning machine for more than an odd day at a time, and the shortage of cans.

The Meddenham Women's Institute, earlier in the field, had managed to acquire a canning machine of their own:

and Mrs Rivers, taking in a stone of gooseberries some time before, had received back thirteen large tins of fruit at a total cost, as far as the canning centre was concerned, of four-and-fivepence ha'-penny. I decided to follow her example with raspberries.

The Manor canes, like the strawberry beds, had recently been renewed and this year produced comparatively few berries. But Walcott, the market gardener over at Fustyweed, had a record crop which he was selling off at sevenpence a pound so as to dispose of them locally and be spared the hazards of transport and the London and Manchester markets. So, having rung up the secretary of the Meddenham canning committee, to make certain that they could be done, and booked a day for them, I ordered a stone of raspberries from Walcotts. Two mornings later, I took them into Meddenham, and then came back to battle with another eight pounds which, thanks to the extra ration a thoughtful Government had lately conceded for the purpose, some parings from my own weekly dole, and part of the small store in the grocery cupboard accumulated by Mrs Dack in pre-rationing days, which Mrs Rivers had turned over to me, I proposed turning into jam.

"How about pickled walnuts, miss?" suggested Mrs Heyhoe, as she busied about the kitchen whilst I stood by the stove and stirred. "The trees are loaded with 'em. An' now's the time."

"Aren't they difficult to do?"

"Not walnuts. Do I hain't never had no trouble with 'em. It's red cabbage as I can't manage. Do what I may, it always go soft. An' all because I pickled my first cabbage

when I wor carrying my first child. An old gypsy woman
told me then as I'd done it, an' that never in all my days
would I be able to pickle cabbage. No more I can't, neither.
I left off trying for ten year, to see if that'd make any odds.
But the next two lots I did turned out soft as a boiled owl,
same as the rest on 'em. With walnuts, now, I hain't never
bin wrong yet."

"What do you do with them?"

"They want to be gathered somewhere about the second
or third week in July. You mustn't leave 'em no later than
that, as a rule, do the nut in the middle'll ha' begun to turn
hard. First you prick 'em well all over. Then you put 'em
into brine . . . a brine just salt enow to swim an egg. They
ha' nine days a' that, changing 'em into a fresh lot a' brine
every three days. Then you tip 'em into a wire sieve. . . .
Croft ha' got the very one in the riding stable . . . an' set
'em out in the sun to dry an' turn black. It may take one
day. Or they might want two. Then you pack 'em into the
jars, cover 'em over with the vinegar what you've boiled up
your spices in, tie 'em down, an' there you are. I'll tell Croft
to get a ladder out of the stack-yard and pull a few, shall I?"

The walnuts were sitting in the sieve on the tennis-lawn,
before I had an opportunity to fetch my canned raspberries
back. Having called in at Cawston's, the engineers, for some
more ploughshares, I turned back into the market place
and stopped Amy outside the Grange, in whose hospitable
kitchen the Meddenham canning was carried out.

The Grange's owner was working in a local military hos-
pital, her family were in one or the other of the services, and
the evacuees who had been billeted there at the beginning

of the war had long since gone back to London, except for
one family who were still occupying the caretaker's cottage
at the back. . . . "And they'll be off, too, the minute no one
else in Meddenham will give them any more credit," the
secretary of the canning committee had told me. I wondered if the moment had already come, for there was no
answer when I hammered on the door. The Grange was
locked. Out in the market place I stopped a passing pram-
pusher and asked if she knew where I was likely to find
anyone with a key.

I was lucky in that I had struck an occasional helper at
the canning. In fact, she had assisted with my raspberries,
she told me: and there they were on the pan.ry shelf, all
ready and waiting for me. But she didn't know what to say
about the key. Had I been to Miss Bailey, down Banwell
Row?

I hadn't the faintest notion of the whereabouts of Banwell
Row. Nor had I the pleasure of Miss Bailey's acquaintance.
Both omissions were remedied within the next twenty minutes, though the long and involved directions given me in
the market place hindered rather than helped in their
achievement. Miss Bailey insisted on my entering her tiny
sitting-room, the walls of which were virtually papered
with photographs, before she would let me state my errand.
And I had to listen to the family history of at least half a
dozen of the sitters, before I could pin her down to the subject of the key. Oh, the Grange key? No, she's never had it,
not at any time. As far as she knew, it hung down at Mrs
Large's, the second of the two thatched cottages agin the
church.

Mrs Large, a minute little person wearing a man's cloth cap skewered on the top of her head with an enormous hat-pin, hadn't had the key at hers for weeks. It was left on a ledge over the top of the Grange back door as far as she knew, unless the Seckerterry had taken it home with her.

By this time I had determined not to go home without my raspberries if I had to drive out to the secretary's home, two miles from Meddenham. Fortunately, when Mrs Large and I arrived at the Grange, we found the lady there.

"How much am I in your debt?" I asked, as we piled the cans on to the wooden fruit tray I had promised to return to the Walcotts.

For answer I was handed a formidable bill.

14 cans of raspberries @ 1/5 = 19 . 10
Credit. 1 stone of fruit 7/- = 7 . -
_____
Balance due   12 . 10

"But Mrs Rivers only paid four and fivepence ha'penny for her gooseberries!" I exclaimed.

"I daresay. But that was when we first started," the secretary began to explain patiently. "Since then the retailers have been complaining that we are underselling them, regardless of the fact that we are canning fruit which would very likely have been wasted in the ordinary way. So the Ministry says we've got to charge the same price per can that you'd have to pay in a shop, less the cost of the fruit."

"Anyway, I paid eight shillings for the raspberries, not seven," I protested.

"We're only allowed to credit the market price, and that

was seven the day you brought them in," returned the canning secretary stolidly.

There was quite a lot I should have liked to say. But this
was neither the time nor the place. So I paid my tribute to
the Great God Profit, and departed.

Vincent ploughed up the Banwell Three Corner, and in
the intervals of carting stones from the Major's gravel pit
to mend the holes and ruts in the Manor roadways, Mark
helped to harrow and roll it down ready for Roger to drill
with mustard. The strip of peas on Banwell Close, less than
half-an-acre, yielded a total of ninety-four bags, exclusive
of the bulging dinner-baskets which most of the pickers
took home with them, and the considerable quantity con-
sumed by me. As the women began to pick the last two
acres on the Upper Prisoners, Vincent moved in with the
'crawler' and the Midtrac plough. Hardly before Barrett's
lorries had carted the last bags away, Roger was redrilling
twelve acres, half with Lincolns, half with Kelvedon Wonder. Within a week the new green shoots were through.

On part of Gunspear and Mautby's Hall the swede and
kale seed, set last summer, had been slowly ripening, and
now it was pronounced fit to cut by a member of the well-
known firm of seed-merchants for whom it had been
grown.

To harvest seed of this kind requires an infinity of care,
for any rough or undue handling will result in quantities
being spilt and lost on the ground. A reaping machine is
useless under such circumstances, so the cutting has to be
done by hand, and soon fourteen of the Manor men set to
work with sickles. The weather was still inclined to be

showery. Occasionally there was a regular downpour. Seldom could the reapers work for more than four hours in a day, and more seldom still for two days together. Before the last of the kale was down, Peel and Catton, having first cleaned out the drum with meticulous care to make certain that no other grains were left inside, took the threshing set up to Gunspear and the swede seed was threshed out straight off the ground, without preliminary stacking.

Peachey and Hill re-horshoed the carrots. Heyhoe and Kettering, with Harry Buxton and Cecil Peachey leading, went and down between the rows of sugar beet, swedes and mangolds with the 'scratchers'. Myhill and Ted Long set off on yet another round of rabbit gassing. Kirk sowed more soot. The young savoys were weeded, and some transplanted on Great Horse Close to fill the spaces left where a coulter of the drill had blocked unnoticed when Roger Marsham was drilling them. Isbell cut weeds and mudded out drains on the upstream meadows. A second crop of hay was cut on Palegate and stacked on the opposite side of the gate to the first. The kale seed on Mautby's Hall was threshed. Four men and a dog (Bowes' black-and-tan mongrel bitch, which is never very far from him) went to mow round the headlands of the oat fields, and Roger Marsham with three horses and a McCormick binder, started cutting down those on Fair Close. Behind him a gang of men stood up the pale yellow sheaves in rows of 'shocks', ten or twelve sheaves to a shock. Soon they were up in the Easter Field. On the first of August Mark and Vincent took the Case tractor and the power binder down the meadow roadway past Croft's and up the New Buildings drift, to begin on the wheat which covered the second half of the Upper Prisoners with a mantle of molten gold. Harvest had begun.

Just before it started Mark and Gidney waylaid Major Rivers in the yard.

In theory a boy first starts work on a farm on leaving school at the age of fourteen, which may be in any season of the year. In practice he almost invariably makes his debut at harvest as a 'holdja' boy, perched astride one of the horses which pull the big harvest wagons, or sitting on the iron seat of a horse-rake. By the time his father, usually an employee of long standing, requests 'a word with the Guv'nor' about his future, he has put two if not three harvests behind him. Mark was approaching the Major on behalf of Frank, his youngest: Gidney spoke up for Evan, his eldest boy. Both had been 'holdja' boys for the last three harvests. (The word 'holdja' is a corruption of 'Hold ye', and is screeched at the top of young voices to halt the horses beside the loaders who are waiting to pitch the sheaves up on to the wagons. No one seemed to have invented a super Mugleston for this as yet.) Now both had passed their fourteenth birthdays, Frank the day the school closed for the holidays, Evan some two months earlier, but both taking the fence together as far as the Board of Education was concerned.

"You can start them both off," the Major said to me when the interviews were over. "Boys are generally more trouble than they're worth. But they've got to start somewhere. And seeing whose they are, they should come in useful for something."

Marsham was not so sure.

"We're got one fourteen-year-old on the books now. An'

you know what they say. One boy is a good boy. Two boys aren't worth more'n half a boy. Three boys, an' you'll get no work done at all."

"Cecil Peachey is more than fourteen, surely?"

"Do he's only just turned fifteen, then. He come last harvest, the week afore Mr Nicholas went. That young Evan, I don't believe he's such a bad un, as boys go. But I aren't so sure about Mister Frank. I reckon he 'ont be a patch on his brother Harry. An' I doubt as he'll never get nowhere near the same street as Mark. They're properly spoilt him, by all accounts, his mother along a' being too soft with him 'cause he's the baby, and Mark by trying to put it right by being too sharp. Bringing up children's like everything else in marriage. You're both got ter pull the same way, do there'll be trouble."

Generally by the time a farm 'goes up' to harvest, the county Agricultural Committee has fixed a lump sum for harvest wages, and it has been agreed between master and men how the work is to be done . . . whether the corn is to be got up 'by the month' (when the farmer gambles on their finishing it in less and having to work the rest of the month for nothing, and the men bet on bad weather prolonging it so they have to receive extra pay) or 'by the harvest', the Manor's invariable custom, and a much more satisfactory arrangement, since there is every incentive to make full use of good weather and get the harvest up in record time.

This year, in Norfolk at all events, no special harvest wage was fixed. Day's work was to be done as usual, and overtime paid for in the ordinary way. On the Manor the only extras were a penny an hour for carting and a small daily bonus for those riding horse rakes, though these were enough to turn the weekly pay book into a species of Chinese puzzle. And while the women remained Patfield's affair, the state of their pay book so amply bore out his oft reiterated plaint that he was 'no scholar', that I sometimes wondered if it would not have been quicker for me to work it out for him in the first place, instead of checking up on it afterwards. Fortunately the current crop of 'holdja' boys were to be paid, as usual, a small lump sum (varying from ten to twenty shillings) at the end of harvest, and meanwhile were booked down only in Marsham's capacious memory.

The old custom of the master's providing free beer for the harvesters has long died out. "Jest as well, praps," murmured Marsham reminiscently. "I can rek'lect some rare doin's when some on 'em ha' had a gallon or two a' beer aboard 'em. Go for one another wi' a pitchfork as soon as look at you, do they got argufying. Though you'd be wholly surprised how much you kin stow away without taking no harm, time you're working out in the fields under a scorching sun."

Nowadays the drink is mostly tea and 'home brew' provided by the harvesters themselves. "Though I don't know how we're agoin' to manage this year, with all this here rationing," mourned Mrs Heyhoe who still spent a certain portion of every day at the Manor, and some presumably at home, but more in running up to whatever field her husband was at work in or pea-picking and helping with the corn on her own account. "Time harvest is on, we'd think nothing a' using three-quarters a' tea an' six pounds a' sugar a week extry, just me an' Heyhoe on our own. To say nothing a' all the sugar I'd want for the malt-and-hop drink I always brew in the copper. An' the men always look to have a hot beef pie or pudding every day time they're harvesting. Heyhoe, he's never missed a day afore, without me bringing him up one in the milk can, with potatoes floating round outside the basin. Now it takes me all my time to manage him one three days a week. A course he can fill up on bread. He ha' to. But I never reckon as that stays with you like meat."

Diffidently I mentioned the merits of milk as a food (though they can obtain it at cost price on the farm, the average family buys no more than a pint a day), the superior staying power of brown bread as opposed to that of the ever popular white, and the numberless corners that can be filled with soup, vegetables, salads, and the like. But the agricultural labourer, be his politics what they may, is a rabid conservative in the matter of food.

"I hain't never bin no lover a' messes an' muck ups. An' we hain't turned rabbits yet. Though there's no tellin' what it'll come to," returned Mrs Heyhoe darkly. "You know, miss, I worn't never one to want more'n my share. But it fares to me all wrong as them as sets about indoors all day should ha' the same amount a' meat as men what ha' got to do a lot a' hard work out in the open. It ought to be divided out diff'rent to that. Stands to reason."

But when has Reason, the passionless goddess, ever been allowed much say in the affairs of men? However, she scored a minor victory in the matter of tea drinking: for a day or two later a paragraph in the local press announced that if farmers applied to the nearest Food Office, giving full particulars of employees, they would be granted a permit for an extra supply of tea and sugar for their canteens.

Possibly there are innumerable farms which run canteens or have facilities for such, though none exist or have been heard of in the neighbourhood of Silford or Ashacre. Still, there was no harm in trying. Stating the total number of men, women and boys at work in the Manor fields, and basing my figures on those supplied by Mrs Heyhoe, I sent in our application to the Sedgeham Food Office. And while the quantity for which I received a permit bore no relation whatever to the amount for which I had optimistically applied, the day arrived when I was able to hand over to everyone on the farm (in an identical pair of packages which I had induced Arthurton's to make up for me) just under a quarter pound of tea, and a fraction over half a pound of sugar. That thanks to the snail-like pace at which officialdom moves, this munificent distribution was unable to take place until the last week of harvest, in no way detracted from its welcome.

The weather was, on the whole, extraordinarily kind. There was a certain amount of rain, well over thirty points on two or three occasions, but most of it conveniently fell at night. During the day it was usually, as Marsham said gleefully, "A wonderful dry time. We aren't hardly having to stop for nothing."

One of the few delays was the barley on the farther half of Booters, which had been cut without tying, and got well soaked the morning of the day before it was due for carting, as it lay on the ground. Next morning we inspected it anx-

iously as the sun continued drying it off. To cart or not to
cart today . . . that was the question.

At first both Marsham and the Major merely muttered to
each other, and looked glum. Then the former brightened.

"Hear that, miss?" he asked me.

I stopped and listened.

A peculiar soft snapping crackling sound was coming up
from the short yellow straw.

"It's adoin' now. Be dry enow to cart fust thing arter dinner, I'll lay a bet."

The Major nodded.

"Yes, you're right. Will can come and set the stack out before he leaves off." His keen grey eyes ranged the field
again once, twice. "You've got a fair crop of barley here."

"That we hev," agreed Marsham. "Most of 'em hain't
done too bad this year. Barring Pound Close. That's werry
moderate. But then that hed a bad start. The rain didn't
come right for it. This here'll help to make up for it,
though."

"Yes. You'll never stack that in twelve yards by five. Will
had better set out a stack and a half."

Roger cut the barley on Gunspear, Crabb's Castle, the Allotments and the Upper Common. Mark and Vincent dealt
successively with Pound Close, Ram's Breck, Banwell
Close, Sidegate Breck and Hunts-and-Gooch. Swath upon
swath, in field after field, was snatched up into the canvas
rollers of the binders and spat out again, tied into neat compact sheaves. Small boys and grown men chased the occasional rabbit with sticks and bloodcurdling yells.

"I ha' know the time when there used to be plenty a' rabbits about, an' every man 'ud have at least one to tak' home

when the field was down," said Marsham as we stood on Three Bridges. "An then the old Guv'nor'd have a mort on 'em to sell up in Norwich an' all. But all this here gassin' an' such ha' done away with all that."

"They must have destroyed a lot of corn," I said, stifling a pang of sympathy for the furry creatures.

"I dessay," said Marsham regretfully. Then suddenly brightened. "There go one."

He chased off after it. Hastily I looked the other way.

The women helped the men to shock up, and in the intervals picked peas on Doles, Crabb's Castle and Barn Breck. The harvest wagons moved slowly over the fields, their progress continually halted by stentorian shouts of 'Holdja', their loads growing at every turn of the blue or crimson wheels. In their wake the horse rakes went up and down, gleaning anything which the loaders had missed or had fallen from a wagon. Stacks rose in one field after another. Walter Bushell and Meale left the carting companies and began thatching. Only the incendiary bomb brooms remained unused and idle where Kettering had stacked them against the fences, for the R.A.F. were reaping a grimmer harvest in the skies.

In July the total of enemy planes brought down over and around Britain had been more than two hundred and forty. August's third week alone accounted for well over twice that number, a hundred and eighty being shot down in a single day. Never in the history of human conflict, as Mr Churchill told the nation, had so much been owed by so many to so few.

The main invasion hadn't happened yet, though it was still looked upon as imminent. The L.D.V.'s, now the

Home Guard, manned their posts as assiduously as ever.
From the French coast enemy guns bombarded Dover.
Bombs continued to fall, especially at 'Random' and on
various 'towns in East Anglia'. The car of a Dennington
man, whose work took him daily into the nearest of the
latter, was to be seen almost any evening in the garage of
the Dennington Kings Head, when anybody was at liberty
to admire the holes made by tracer bullets. Men at work in
the fields and round about farm buildings were fired. A
passing raider machine-gunned (without effect) the Ashacre Hall herd of cows. Major Rivers, who had been carefully going into the question of income tax since the appearance of the new budget, retaliated by making arrangements for rendering one of the rooms on the ground floor at
the Hall more or less bomb proof, at great expense, and ordered a concrete shelter to be built into the ground at the
entrance to his drive.

Whilst they were under construction, he turned his attention to the Manor.

Accompanied by the contractor, a large and imposing
gentleman whose considerable dealings in A.R.P. had already endowed him with a most sumptuous car, a formidable bed-side manner, and a professionally pessimistic attitude as to the outcome of the war, he marched from room
to room. Finally they decided between them that the cellar
was the only possible place to tackle, and that this should be
undertaken without delay.

It should be supported with steel girders and its roof reinforced in such a way that the entire house could collapse
on it without a scratch on anyone inside, the contractor in-
toned impressively.

Two exits, one the inside-stairs leading down from the back hall, the other through the grating on to the drive, which could have a manhole cover on top of it, and a ladder inside, the Major cried enthusiastically.

Personally I would much rather be blown to pieces in the open air than die like a rat in a trap.

"What about fire? Or suffocation, if you're not dug out in time?" I asked sceptically.

Both looked at me like two misunderstood children confronted by a tiresome grownup who threatens to take away their toys. It seemed a shame to interfere with them. Anyway, it wasn't my house or my money. So I left them happily at it, and went up Church Lane to see how the stack on Ram's Breck was getting on.

When that was finished, there were only two more fields to cart.

On the Manor farm, dwarfing such happenings as the first air-raids on London, Eire, and Berlin . . . what were air-raids, anyway, to veterans like us? . . . the main pre-occupation was whether or no we should manage to get the harvest up in August, a feat which is seldom achieved. Certainly we should be the first to finish in Silford and Ash-acre if we did, though the privilege of parading through the district blowing triumphant blasts on the harvest horn is a thing of the past.

"My father used to tell me as they had a proper horn in his time. An' them as finished fust ud be off and round all the fields a' them as hadn't, blowing it fit to bust. In my young days we used to cut a piece a' elder an' take the pith out, an' blow down a' that. A rare row we used ter make, a' shouting an' singing an' all. An' finish off by randying at the Hart. Then let a day or two pass, an' the old Guv'nor ud give us a supper. Beef, boiled an' roast. Plum puddings. All the beer you could swaller. An' baccy for the asking. All set out as nice as pie in the corn barn. An' a social afterwards, what our old women could come to, and we wouldn't break up till morning come."

"A pity you've let things like that die out."

"Ar. We hain't hed a do a' that sort since Mr Nicholas turned twenty-one." For a moment Marsham's face clouded.

"If we finish in time, you'll have to cut me an elder bough to blow. And I'll stand you a 'Norfolk Nip'," I promised him.

The thirty-first of August fell on a Saturday, the normal week's half day. But this Saturday no one was looking at his watch at one o'clock.

Seven and a half hours later Marsham solemnly presented me with a hollowed piece of elder bough some twelve inches long, and nearly as thick as my wrist. As solemnly I accepted it, raised it to my lips and blew. The result was nothing but a muffled squeak, like the cry of a new born kitten.

"No. That ain't right."

Marsham snatched it back again.

"This here's the way."

He put it to his mouth.

Across the Upper Common there shrilled a long clear piercing note.

❧ IX ❦

"IT'S wholly warm," Mrs Heyhoe said thoughtfully, as she was preparing to mount her bicycle and make for home at mid-day. "Do it get out like this tomorrow, I don't know as it wouldn't pay me to leave off my long-sleeved winter vest afore Heyhoe and I catch the bus into Norwich."

I stared open mouthed. In a linen frock and very little else, I was nearly melting on this September day with the thermometer ninety-one in the shade.

"I never like to leave things off in too much of a hurry," she went on, as she caught sight of my face. "I take after me feyther. He always went to work in three coats, winter and summer. He said he reckoned as what ud keep the cold out ud keep the heat out. An' I don't know as he wor so far out at that."

Mounting, she rode cheerfully away.

I went on into the house and down the cellar steps to see how the air raid shelter was progressing.

Here at least it was cool, though the two men who had been at work in it for the last five days said they were warm enough.

"Did you hear the bombs yesterday evening, miss? They

were wholly lumping 'em down when we was on our way home." Alf, the smaller of the two, was spokesman as usual. 'Chicken', his mate, seldom did more than second his conversational efforts with an expressive grin or an occasional 'ar'. "Me an' Chicken saw a Spitfire arter 'em as we were going along the Bull road. There worn't half a shindy. Chicken, he would get off his bike and set in the ditch. But I stood on the fence and watched 'em. It was a do."

"Ar," remarked Chicken sourly.

"Well, what I ses is, if a bomb's got your name on it, it'll chase you round corners. If it hain't, where's the use a' worrying?" returned Alf. "They dropped one on the train my brother come home on from Yarmouth the other day. And if it had had another carriage they'd ha' hit it. As it wor, they only knocked a bit a' glass out a' the windows. And nobody was hu't."

"Ar," said Chicken noncommittally.

"You're gotter laugh," said Alf. "You should ha' heard me an' the missus the other week. The sireen had gone for the all clear an' we'd just got the kids out a' the shelter an' all gone back to bed, when down come some more, just at the back on us. The missus, she got blown clean out a' bed an' on to the landing. An' there was I, sitting in the fireplace covered with soot. Laugh. I thought we'd ha' died. 'You want to see your face', says she. 'You don't look no oil painting yourself, come to that', I told her, seeing the mess she was in, with the plaster from the ceiling. But we soon got cleaned up again. Now some folks don't know when they're well off. I told you about that job with the school teacher the other day didn't I?"

"No. I don't think so."

"The missus, she'd gone down into the city to get the schoolmaster to sign a paper, so she stood there an' see it all. The warning went, an over come jerry, machine gunning an' going on. The children were as good as gold. Marched off down into the shelter the minute the teachers give the signal, as nice as pie. All bar one youngster, what stood there stare-gawping just outside the dugout, with bullets a-flying round. As she dived in, the schoolteacher, she had him by the seat a' his trousis, an' pushed him in a' front a' her. Saved his life, I reckon, an' all. An' what do his mother do the next-day morning but come round complainin' an' making out as she ain't going to send him to school no more, 'cause his trousis got tore. Then you talk about it. But there, you're gotter laugh."

Alf was always ready for a gossip, though he never wasted any time when at it, but worked while he talked.

"Don't suit me to stand about with idle hands," he told me. "I ha' bin unemployed, so I know what I'm talking about."

"For long?"

"Long enough." For a second Alf's philosophy sagged, and he scowled. "Twenty-seven-an'-sixpence a week was what I drawed time I was out. An' that ain't a sight with a wife an' three kids an' the rent to pay. What made it worse was along a' its being summer, when we'd looked to go on some a' the excursions to Yarmouth. Howsumever, we got round that all right." The scowl vanished. Alf was himself again. "We borrered a couple a' push-bikes, an' went all round the coast, each on us with one a' the kids on the carrier, an' having the baby in front turn an' turn about. We took our grub along with us. An' the missus, she come out with half-a-crown she'd contrived to save up, so we took that along with us an' all. But we never broke into it. Every day for a fortnight we see somewhere different. It wor the best holiday as we ever struck. But I wor somethin' glad when I got a start agen. It ain't much of a job, loafing about, arter the first day or two. What do you say, Chicken?"

"Ar," replied Chicken weightily.

The two were keen shelter connoisseurs, having assisted in the building of a considerable number since the war began, as well as having perforce spent a certain amount of time taking refuge in them. Neither thought a great deal of those in the 'East Anglian town' in which they lived. ("Don't go down far enough," said Alf, who must have been a mole in a previous incarnation.) The Manor cellar ranked as 'not so dusty', though they allowed that there was just the odd chance that I might be right about getting trapped in it.

"Pity our Guv'nor couldn't ha' got you that steel door," admitted Alf, eyeing the arch at the foot of the steps. "That 'ud ha' shut you off proper do they wor to land one atop a' them backstairs. But don't you fret, miss. They could keep on hulling 'em down for a month a' Sundays without contriving to get one just there. An' do they was to, that concrete baffling 'ud stop a master lot a' blast. An' that'll take a funny old bomb to shift them girders. I reckon you 'ont take much harm down here."

The pride of their hearts, however, was a dug-out they had recently helped to make at a semi-stately home on the other side of Meddenham: ten men had laboured for six weeks at this masterpiece. Built in the garden and well away from the house, it could be reached either from outside by a short tunnel whose entrance was screened by a clump of laurels, or from inside the house via an underground passage which ran below the drive and beneath a long flower bed bordering a lawn. The actual shelter, measuring sixteen feet by twelve, had an inner roof of steel upheld by steel girders, a secondary roof of reinforced concrete three feet thick, and finally several feet of earth crowned by an outside rockery.

"The sky could topple down on it an' you'd still be there. Yet looking at it from the outside, you'd never know as there was nothing there," cried Alf ecstatically.

"Ar," corroborated Chicken with enthusiasm. "That one we made on that farm out Sedgeham way warn't too bad, though," he went on unexpectedly.

"What, that one a' Mister Syderstone's? Oh, ar. Seems as though the old gentleman had always been meaning to build a proper place underground to store apples an' potaters in. Been talking about one for years, so his missus told us. But that was as far as he got. Then they come an' dropped a few bombs alongside his cowshed, so he thought as he'd get a move on with it. We fitted it up properly time we was about it. They're going to sleep on the shelves time the war lasts. And then, when it's over an' done with, on go the potaters an' the apples."

"Talking of apples. . . ." For days past the Manor orchard and kitchen garden had been littered with them. The men, invited to help themselves, had grown tired of pick-

ing them up. Besides countless chip baskets full of plums
and damsons (as Peacock had foretold, there was a record
crop of 'stone fruit' . . . even the hedge behind the pig
houses was red with small sweet wild plums) I had taken
sackfuls of windfalls round in Amy to the troops quar-
tered at West Malton Hall. I had posted boxes to London.
Still the apples fell. One of the tragedies of today is the way
in which Nature's bounties are so often wasted, and with-
held from many who would be glad of them for lack of
adequate means of distribution, or because 'it wouldn't
pay.' But perhaps in this instance Alf and Chicken could do
something about it. Would they care for some apples for
themselves? And did they know of any others who might
like some?

Did they not!

"Hain't seen an apple since I don't know when. No more
hain't anyone else round our way. An' I was always one for
an apple tart. What about you, Chick?"

"Ar," responded Chicken, with a genteel smacking of his
lips.

For once at least, scarcely a windfall rotted on the ground
at the Manor.

The price of peas dropped from between three-and-six
and five shillings a bag to a shilling, even sixpence . . . and
this was without commission deducted, to say nothing of
such incidentals as transport, labour, and general produc-
tion costs. So the Upper Prisoners crop was harvested, carted
home, and stacked on the edge of the stackyard with the
last remnants of Barn Breck. Mark, Vincent, and Leslie
Long began to plough cultivate and riffle up the stubbles.

Isbell went back to mudding out the drains. The new re-
cruits were set to pulling 'quicks' (the local name for
docks), with Marsham keeping on them a jaundiced and
critical eye. Finding the muck-carters leaving off at half-
past three one afternoon, I learnt that on the Manor farm
this is always worked by the 'stint' . . . twenty loads on the
old fashioned tumbrils or fifteen if they are using either of
the two large iron ones . . . and once a man has finished
his 'stint' he is at liberty to go home or claim overtime.
Hedges were trimmed ('browing' is the correct term). The
roadway ruts were filled in. The large-scale 'blitz' on big
towns began, with London as the first victim.

Mrs Rivers had enlisted me as W.V.S. representative for
Silford within a week of my arriving at the Manor, but so
far my activities had been confined chiefly to organizing
sets of gas lectures, issuing knitting wool and collecting the
resulting 'comforts', collecting scrap-iron for the Ministry
of Supply and securing billeting allowances for self-evacu-
ated families who descended on relations in the village
from time to time. At intervals we were warned that we
must be prepared to deal at any time with a large influx of
refugees at twenty-four hours' notice, but while the unofficial evacuees continued to flow into the neighbourhood
from London, Liverpool, Birmingham and sundry East
Anglian towns, the official flood was held up. The 'blitz'
on London opened the evacuation sluice-gates with a rush.

I had planned to spend a weekend in mid-September
with M., who had recently removed herself and the children to a cottage in Essex. But on the Friday morning Mrs
Rivers telephoned to ask if I could postpone it until the following week.

"It's the bombed-out families. The Sedgeham W.V.S. people have just rung up to say they're coming tomorrow or Sunday."

"How many will there be?"

"Nobody seems to have the slightest idea. But so far Ashacre and Silford are only down for two families each. And they say that no family will consist of more than a mother and two under-fives. So will you give the cottages the once over, and stand by?"

When, just over a year ago, the Government had launched the first evacuation, it had looked little farther than the problem of transporting the evacuees from one place to another. In those early days, too, the country folk, carried away on the high tide of self-sacrifice and heroism which usually prevails at the beginning of a war, had been ready to throw open their hearts, their purses, and their homes, without reserve. Since then the countryside had, to put it crudely, had some: and not unnaturally, was decidedly averse to a second dose of the mixture-as-before.

"There's only room for one woman in this house, and that's me," I had been told, over and over again, when I had taken a house-to-house census of possible billets, soon after Dunkirk. And "Why should I knock myself up looking after other folks' children, when I've got all I can do to see after my own?" Or "How can I go out to work in the fields, leaving a stranger at home? There's no telling what she'd be up to."

Then there are the pretty set of posers presented by rural sanitation (or lack of it), and the water supply. In Silford as in other villages, the latter depends on wooden barrels set under cottage eaves to catch the rain, and a few wells, one or other of which is always threatening to run dry. The former question is still more vexed. "We ha' hed to dig over most a' the garden as it is, 'cause the night-cart don't come our way. Do we was to hev any more in the house, I don't know where we'd be," said Mrs Marsham, speaking not only for herself but half the parish.

And while no one in Silford had yet given house room to a Government-sent evacuee, everyone knew someone who had, and proceeded to enlarge on it.

Mrs Mears at Dennington, hers were 'alive' when they come, an' her own three had got nits off of 'em, as well as the measles.

They'd sent two fourteen-year-old girls to old Miss Blyth, down at Pennyspot, and first thing they said to her, soon as they'd had their tea, was "Any boys round here?" She couldn't do nothing with 'em at all.

Mrs Elscey at Wendeswell, she'd had to turn hers out in the finish, an' then burn all the matting an' the bedclothes. She's disinfected the place out an' all, but you could still smell 'em the minute you put your nose inside of the door. She say they can put her in prison afore she'll take any more.

Mrs Tye, on Banwell Common, she'd had a red label. But the baby worn't due for well over a fivemonth, an' there she set, expecting to be waited on hand and foot, and when Mrs Tye, she spoke to her about it, all she got was, "The Government's paying you eight bob a week to look arter me. Do you get on with it." An' Mrs Guymer, what lived next door to Mrs Tye, she'd had to have hers up in front of the Bench. A mother an' daughter, she had. It's right the

daughter worn't much. Nothin' more'n a substitute off the
streets. But the mother, she worn't much better, and Mrs
Guymer, she got sick a' seeing her knock the girl about.
But when she come to interfere, the mother set about her
an' all.

Several householders had 'missed' things. Others had had
cherished possessions wantonly damaged or destroyed. The
first question asked of quite a few hostesses was, "What
time do the pubs open?" which put the visitors beyond the
pale at the start, apart from the well publicised incident at
Sunnington where the police had had to be called in to deal
first with ten hilarious evacuee mothers who refused to
leave the 'Green Man', and then with ten indignant hostesses who refused to readmit them. For while in Bermond-
sey or Rotherhythe, wives as well as husbands indulge in a
companionable glass, in Silford and such, the 'local' is a
purely male preserve, and it is a bold woman who ventures
her reputation by joining them any evening.

But even with the many evacuees who were neither shift-
less nor in search of a holiday at someone else's expense,
there were . . . and are . . . countless evacuation problems to which Whitehall has apparently never given a
thought. And still more which they can do nothing to solve
if they have, short of scrapping the whole patchwork
scheme and starting afresh by building self-contained
camps, small townlets which could be run on communal
lines, and to which whole communities could be evacuated
more or less en bloc.

Though from time to time he will succumb to one ism or
another, and for a while function comparatively success-

fully as a group, at bottom the human animal is an incurable individualist. For a single family of any size or character to share a house in any kind of harmony is undertaking enough. When there are two, neither with much choice in the matter and both with entirely different backgrounds (and with the best will in the world, country folk regard 'townees' as an odd and unaccountable race, and vice versa) the possibilities for friction are endless.

Nor is the animal kingdom exempt. When West Malton Hall made its hurried departure to Cornwall, the matron had begged me to give temporary hospitality to her cat. Chips was not a particularly engaging creature. Still, he was harmless enough, and ready to be properly subservient to Samuel Pepys. But after a month of internecine warfare, the best they could achieve was a species of armed neutrality . . . both slinking about the house, peering furtively around corners and under furniture, and eating meals with one eye always on the door . . . with hostilities liable to recommence at any moment. I was as relieved as they must have been when eventually Chips' owner wrote for him, and I packed him into a hamper and put him on the train at Dennington.

Mrs Heyhoe put the case for all parties one morning when, terrifyingly and unexpectedly, she burst into tears all over me in the middle of a discussion on the best way to make apple rings.

"I reckon as I'll go clean off my dot if me sister-in-law don't take herself an' Ireen an' Margy off soon," she wept hopelessly. "There they are, all the time, an' we never fare to be on our own at all. Clara, she don't fare to settle, neither. She keeps fretting after Jim an' the place she come from, an' says the country's too quiet, an' there's nothing to do. There's plenty to do, but she 'ont get on to it, not being used to our ways. An' do you say anything to the children, an' you know what children are, always up to something, she rear up at once. I can't hardly open me mouth, without there's someone jumping down me t'roat. Heyhoe an' me, we didn't never used to have no words. Now this week he 'ont hardly talk to me. He will have it as it's along a' them being his people. An' do that had ha' bin my sister from Battersea, I'd never ha' made a dean, he reckon. But it ain't nothing a' the kind. It's never having yourself to yourself, an' not knowing how long it's going on for."

"I'll ha' them do I'm forced. But not until," was the general verdict, which Mrs Rivers had frankly endorsed.

"I don't want to be mean. But really, I've got as much on as I can cope with just now," she said. "And it seems a shame to let poor old Nick's place be mucked up while he's in a German prison camp, unless it can't be helped. Charles says we shall have to fill up with landgirls before we've done, anyway. If there's a deluge, of course, we shall both have to have some. But for the time being we'll hunt up any empty cottages there are, and see what we can do with them."

In Silford three such had come to light, one down at the 'Hole', recently vacated by an old age pensioner who had removed to the home of a married daughter in Wendeswell; and a pair in Port Row, a turning leading out of what passes for Silford's main street, which had been untenanted for the last eighteen months owing to the owner's unwillingness or inability to pay for certain necessary repairs. These were soon commandeered by 'Miss Kate', the repairs to those in Port Row promptly put in hand, and the rooms of all three scrubbed out by half-a-dozen willing helpers. The Ashacre schoolmaster sent his pupils round to collect Danegeld in the shape of promises to give food, furniture or cash, if and when the evacuees should come.

It now remained for me to see that these were implemented. So, having wired to M. to expect me tomorrow week instead, thrust the pay envelopes at Marsham, and fetched blankets and palliasses from Sedgeham, I hitched the trailer to the back of Amy and set off to pick up such goods and chattels as the donors were unable to transport to Port Row by themselves . . . a considerable quantity over and above those on the original list, since everyone was ready to express his sympathy in as practical a way as possible. It was only the throwing open of their homes at which they drew the line.

Since time was so limited, and not more than two grown-ups and four children expected at the outside, it seemed best to concentrate on a single cottage. Probably the strangers would be glad of company for the first few nights. They might, perhaps, decide to join forces. Quite possibly, the Sedgeham depot warned me, they would catch the first train home again on Monday morning. At any rate one cottage should fill the bill for a day or two.

By Saturday noon everything was ready.

Peacock had repaired the windowpanes on which local talent had demonstrated its ability to throw a straight stone since my last visit. Two cwt. of coal and a month's supply of kindling were stored outside in the 'copper house'. Beds were made on palliasses in the two upstairs rooms, with bedsteads promised for next week. Downstairs the larder shelves bulged with food. Those in the kitchen held crockery and saucepans. Brooms stood in a corner. Half a cwt. each of carrots and potatoes were stacked underneath the sink. In the livingroom a fire burned brightly. On the hob sang an aluminum kettle which had arrived too late for the recent Spitfire collection. There were pictures on the walls and a looking glass over the mantelpiece. On the table a vase of flowers pinned down the sheet of foolscap on which was listed a number of things I must remember to tell the visitors without fail. This included such items as an injunction always to use the new pail in the larder for drawing water from the well, to empty slops in the farther garden as the first was let to Patfield, and the news that Mrs Jordan at Yew Tree Farm would send in their Sunday dinner tomorrow. The score or so of children, for whom the task of finding errands to run had taken almost as much time as furnishing the cottage, had been persuaded to go home. And Mrs Peel, who lived just across the road, had promised to look after the fire, and put up the blackout if perchance dusk came before the evacuees.

The bus which was to bring them out from Sedgeham was to deposit the evacuees at the Manor, where I was to give them tea and then take them up to Port Row afterwards in Amy.

At half past two Mrs Rivers rang up to say that I could expect them at about four o'clock. At five she rang again to say it would probably be half past six. At half past eight

I rang up Sedgeham and was told that so far only half the expected number of evacuees had turned up, and some of those were still to be sorted out. At ten o'clock Sedgeham rang me to announce that no more would come in tonight. A few were still on their way to billets, but whether or no any of these were consigned to Silford, the speaker could not be sure. If none arrived within the next half hour, I could take it mine wouldn't come until tomorrow. Sunday and Monday followed a somewhat similar course. On Thursday Sedgeham reported that the next batch of evacuees would arrive on Friday, when Silford's would come for certain, and I sent a second wire to M. A week later, having endured a repeat performance, I despatched a third.

Bushell and Meale finished thatching, and armed with a pointed stick and a steel measuring tape, I went round all the stacks with the former to 'measure up' before the final payment was made. Thatching is done on piece-work, and is paid for by the yard, the yardage being ascertained by plunging a stick in at one end, at right angles to the stack, and noting down the length of one side. Bushell had already presented his total, having 'measured up' with his own tape as he went along, and was none too pleased at my being made to check up behind him.

"If the Guv'nor don't trust me . . ." he began aggrievedly.

"Nothing of the kind. He wants me to learn how it should be done," I interrupted hastily; though in fact both Major Rivers and Marsham had cast aspersions, if not on Bushell's honesty, at least on his arithmetic, when they first heard the result of the measuring up.

"We used to have a rare time with his feyther, when he worked for the old Guv'nor," the latter elaborated. "Kep' pigs, he did, an' all. An' do the guv'nor han't a' kep' an eye on him, he'd a med off with enow hoss corn an' meal so he'd never ha' gone nigh the mill on his own account. Master fond a' walnuts he wor, too. The old Guv'nor, he never said no to any on us what asked if we could pick up a few. But Flip, he weren't content with that. What must he do but t'un up early one morning an' brung a sack. I see him under the trees, a muttering away to hisself same as he always did, though I never let him see me. An' I watched where he put 'em under a heap a' chaff in the haybarn. Then I hed 'em out an' shared 'em round among the rest on 'em. Just afore leaving off time, I hid up in the barn, an' waited for him ter come along an' look for 'em. He wor wholly a' mobbin' away to hisself time he poked about in the chaff. I hed ter smile."

In this instance, however, they were, as Marsham handsomely admitted, barking up the wrong tree, for my total exceeded Bushell's by some five yards, a matter of over eleven shillings.

Mushrooms poked up through the stubble on the top of Crabb's Castle. The hedgerows turned russet, crimson and purple. Coils of rusting dannert wire hung, golden-brown, across the brambles beside clusters of bryony berries, scarlet, orange, and green. The last of the swallows and housemartins held mass-meetings on the electric current and telephone wires, before setting out in search of the sun again. Walnuts rained down from the trees on the Walnut Meadow. The evenings were drawing in. In the mornings the hedges were hung with dew-spangled cobwebs. There was a nip in the air. From over the shoulder of autumn the countryside looked on the approaching face of winter.

In my ignorance I had always thought of this season of the year as a slowing down, a time in which the land pauses to draw breath before it makes the effort needed in Spring. I had seen myself, too, with time on my hands, now that nearly a year's experience lay behind me. But spare time on a farm is apt to be as elusive as the jam in the proverb. Jam yesterday (only somehow you never saw it), jam tomorrow, but never jam today. No season is ever like the one which went before. The weather creates fresh problems every day. And at the moment, whenever the weather left off, war conditions began. While far from leaving me freer, experience merely underlined the truth of Major Rivers' earlier remarks about farm pupils, and made me realise how very little, in spite of my frenzied busyness, I had accomplished so far.

Four cows, sentenced to fattening some weeks back, were struck off the record sheet and despatched to Meddenham market. The fifteen in-calf heifers which had been summering on Three Bridges meadow, were brought up into the herd, and as each one gave birth to her first calf, she was earmarked, numbered, named, and allotted a page in the record book . . . the first three featuring as Spitfire, Blitzkrieg, and Hurricane.

Repeated official warnings had arrived by post to remind us that during the coming year farms were expected to be practically self-supporting in the matter of feeding stuffs for stock, so Gypsies, as a wheat stubble, was scheduled for beans for cows. The manure from the east yard at Silford Hill was carted out and spread over the field. A small box-like affair . . . the bean drill . . . was bolted on to the double-furrow plough. As Vincent, on the Case, ploughed up the stubble, Jimmy Bailey, perched uncomfortably on the plough, worked the drill, and ploughing and bean-sowing were performed in a single operation. Furrow after furrow was neatly turned, the beans scattered in one round being covered over on the next. Behind them Mark with the Fordson cambridge-rolled, then disc-harrowed, first following the line of the furrows: then, when Vincent had finished, 'over-wart', or across the furrows, until the heavy stubborn soil of Gypsies was sufficiently fine sown.

Sugar beet lifting was started on the North Field. The Major rang up to say that the recent high winds had torn loose a number of London's barrage balloons, some of which were now drifting over Norfolk, and would we keep a sharp look-out . . . which everyone enthusiastically but unavailingly did. The weather turned cool enough to begin working carrots, so Peachey ploughed out an acre or two and the women started picking on the Limekiln Breck. Nemesis, in the guise of the War Agricultural Committee, descended on Frederick Lophan, dispossessing him of the glebe land, and requesting Major Rivers to farm it with the Manor: so Myhill and Harry Buxton went off gassing rabbits, and Vincent started in the field adjoining Coney Hill with the 'crawler' and the 'digger', ploughing sixteen inches deep in the hope of burying up some of the foul grass. The sugar beet lifters moved over into Swains; and on the North Field most of the 'tops' were carted off and thrown about on Ling Close and the Pig Pasture for the cows, the few remaining being hastily spread about the field so that Leslie could begin to plough it up.

"Whatever you do, keep the ploughing going for every minute the weather will let you," the Major never ceased to impress on me. "It makes all the difference to your crop if you can get the land ploughed over before Christmas, as I told you before. Those hilly bits on Crabb's Castle will have to be done with horses."

The importance of keeping every horse occupied was also well drilled into me.

"Marsham's a grand chap. I'm not saying anything against him. But you have to keep an eye on him, just as you must on everyone else. Only yesterday afternoon I happened to look in the stable, and there stood a horse, doing nothing," the Major, who seemed to have eyes everywhere, told me one morning. "Tractors can stand idle. But if there's one thing you mustn't do, it's to let horses stand about in the stables, eating their heads off. And that's one of the little things where Marsham trips up. You mustn't leave it to him, but go into it with him overnight and work out a proper time-table for the following day. He's much too fond of letting the horses take it easy, and running Mark about instead, wasting fuel and wearing the engine out."

In theory, with thirteen working horses, as there were at the moment, there should have been no difficulty in managing without undue calls on Mark. In practice, I was forever working out complicated sums which refused to add up to the correct total.

"One for Peachey ploughing out carrots. Three for Heyhoe on Crabb's Castle," I would begin. "One for Roger drilling, and another for Kirk sowing the basic slag. One for the cowmen and six carting carrots down to the washer. That'll do, won't it?"

"And what about rolling down behind Leslie on Booters?"

"Must that be done tomorrow?"

"Yes, do there come some more rain, you 'ont be able to touch it. An' Tom, he must ha' some straw for the cattle at Silford Hill. Shud ha' had it last week by rights. He can't go no longer."

"Manchester only wants five tons tomorrow, so Peachey can finish in the morning and cart in the afternoon. That'll give us another one. And what about the cows? Can't the boy cart enough in for them before he leaves off for dinner?"

"Do he'll never get the tops off Swains in front of Vincent ploughing."

"It looks like rain, anyway. If it does, that'll stop Heyhoe and Roger and Kirk, and give you five horses."

"Then they'll ha' to go muck carting," says Marsham relentlessly. "And who's to cart off the sugar beet?"

I give in.

"It looks as though Mark will have to cart some carrots in to the washer before he takes the Manchester ones in to Westham. And Bushell had better take the other Fordson and cart two loads when he's finished in the washer. But you know what the Major will say."

And, of course, he never failed to say it.

The carrots were generally the villains of the piece as far

as transport was concerned. But the difficulties they made inside the Manor boundaries were nothing compared to those which had to be dealt with outside.

"We can't take any more carrots up to London at the moment," announced Barrett's after the first load or two. "They've cut down our petrol again. And they're making no end of a fuss about road transport. You'll have to get a permit before we can do it again, in any case. And then it probably won't pay you. We can't get through at night now because of the blitz and the blackout. They have to go up next morning, which means they don't catch the market until the day after. You'd be much better off on rail."

"We're doing our best," said the Westham station-master in an injured tone, when I rang up for the nth time to report delays in delivery, and more bags of carrots lost in transit. "But there's been some more bombs somewhere up the line. And were you thinking of sending anything up to Spitalfields tomorrow? Because they tell me it's just been hit, and nothing can get in."

"Many thanks for your letter," replied R. A. Leeds. "What has happened to Spitalfields? Absolutely nothing at all. We are completely at a loss to understand the Railway Company advising you in such a way. We have been carrying on normally and doing quite as big a volume of business as usual all the way through. It is true that at times delayed-action bombs have been around us, but nothing has interfered with our trade until yesterday, when owing to comparatively close proximity of two Land Mines and an unexploded bomb, the Market was evacuated from mid-day onwards, but we resumed normal trade again this morning."

"All the same, I think you'd better cut out London for a time," observed Major Rivers, when I showed him this last. "Look at the returns this morning. That sort of price is no earthly use. The carrots might just as well stay in the ground. I don't believe things will improve much, either. We're going to have our work cut out to get rid of the carrots this year, if you ask me, in spite of the way they keep on bleating about a food shortage. Manchester will take a few more. But they like the large ones, not the London size. And they never give much of a price for them at the best of times. I think you'll have to take on that canning job, after all, though there isn't a lot hanging to that."

"They're paying the carriage. And they want them dirty, which will save all the washer expenses," I reminded him.

"I know all about that. But you'll find you'll have to give extra weight to allow for the dirt before you're done. And then they'll be forever complaining. And what size do they want?"

"Not more than an inch and a half, and not less than half an inch across the top. Patfield says that will mean an extra ton an acre, as London won't take them as small as that."

"So it may. And a nice job you'll have keeping the women up to it," prophesied the Major. "They'll have to pick into three skeps at once . . . Canning, London, and Manchester. And the London ones will have to be mixed off with Manchester's down at the washer as best they can. You'll have to choose your spots too. This year's carrots are nothing like as big as last. But you've got some pretty decent ones. You won't find any canners on the Pightles, for instance. And only a few odd patches on Prisoners Close. There are

several on Limekiln Breck, but they're all over the place.
You'll have to dodge about from one field to another."

"There are three or four acres of smallish ones at the bottom of Doles."

"I know there are. But they're all on twenty-inch work.
Roger can mould those up with the Northumberland
plough. The eighteen inch will only take an ordinary one,
which means they won't stand up to so much frost, and
they're the ones you get rid of first. Whereas Doles
shouldn't take much harm until next April. As a matter of
fact, I had an offer for the field more or less as it stands the
other day. A chap who's got hold of some Government contract or other, and wants them dirty, free on rail, in March.
He hasn't got up to my price yet. But I think he will before
he's done. That and the canning people should clear the
decks a bit."

The canning company signified that they had much
pleasure in confirming the purchase of small carrots, that
they would be glad to receive five tons a day until further
notice, and that they were railing sacks immediately. The
Dennington station-master, when interviewed, undertook
to do his best about trucks. The District Goods and Pas-
senger Manager, approached by telephone, promised to
look into the matter of 'Service'. While we waited for the
aforementioned sacks to arrive, we loaded a few hundred
bags of carrots to London and a steady forty tons a week to
Manchester. Roger Marsham drilled wheat on the North
Field, Lincolns, Gunspear and Palegate. Harry Buxton and
Sidney Peachey who had, according to Marsham, "bin most
oneasy-like," ever since harvest, suddenly handed in their
notices and left to better themselves as builder's labourers at

a new aerodrome in course of erection nearby. Major Rivers bought two yearling Percheron colts, which were turned off on the Walnut Meadow with Kitty's foal. An outbreak of foot-and-mouth disease at Wendeswell, and all the victim's stock with a cloven hoof had to be slaughtered, even the pet lamb beloved by the farmer's small daughter. In a couple of days the work of a dozen years and one of the best herds of milch cows in the county, were destroyed. All movement of cattle within a fifteen-mile radius was prohibited. Since we were within the two-mile limit, both the Manor entrances were strewn with straw well soaked in disinfectant which was renewed every day, for the benefit of wheeled traffic. The carrot pickers approached Limekiln Breck from the main road and were forbidden to come through the yard. Anyone entering the cowhouses had to soak his boots in the pails of disinfectant which stood by every door. In addition the Manor cows had to be inspected night and morning for symptoms, with Lambert reiterating afresh on each occasion his considered opinion that the outbreak was the result of enemy action. At intervals I rang up the Banwell station-master and persuaded him to use the station wire to Dennington (which was not, alas! on the telephone) to find out if the canning company's sacks were in yet. For the fifth weekend in succession, the Sedgeham W.V.S. announced the imminent arrival of the bombed-out families, and the Port Row cottage was re-swept and garnished . . . a trifle perfunctorily this time.

"Do they don't come soon, we'll ha' used up all their firing," remarked Mrs Heyhoe dispassionately, as we banked up the fire before leaving on Saturday afternoon.

The remainder of Saturday, and Sunday morning and

afternoon ran their now familiar course. At half past six, just as I was about to begin another apologetic letter to M., I became aware of some muffled rumbling noises coming from the other side of the house. As I reached the front door step, a large red bus was pulling up in the drive: and as it braked to a standstill, a harassed-looking official waving a note-book sprang down from his seat beside the driver.

"Is this Silford Manor? Yes?" He consulted his note-book, then put his head inside the bus. "Two families, please. Out you get."

Two youngish women, a toddler, a couple of small girls who might have been five-year-old twins but turned out to be sisters of six and ten respectively, a push-chair, three suit-cases, and a miscellaneous assortment of paper parcels and carrier bags, were hastily decanted on to the drive.

"Sign for them, please," the official said briskly. I signed. "And now how do we get to Ashacre Hall?" He was already back in his seat.

"Turn right by the sandbagged post at the entrance, over the bridge at the bottom of the hill, left by the concrete pyramids, and follow the road for another two miles, when you'll find the drive on your right, with a barbed-wire road block and some old wagons on the corner."

The bus backed round, churning up the gravel (Croft would have something to say when he saw it), then lumbered off down the drive, shredding branches from almost every tree as it went. My visitors dumped their luggage into the waiting trailer and followed me indoors for tea, the elder of the small girls collecting a protesting Samuel Pepys en route.

By the time tea was over I had learned a good deal about them beside the information I needed, as billeting officer, to fill in the inevitable forms.

Mrs Waggett, resplendent in navy slacks, an orange jumper, a turban and a tweed coat, was an old hand, this being her fourth evacuation. The last time she'd fancied Devon, and had spent the summer there. But towards the end of August, she'd said to herself it was time she had another look at London, and gone home . . . only to be bombed out in less than three weeks. So here she was, off again. There was one thing about it, you did see round the country.

Mrs Stubbings, parent of Sheila and Margaret, lacked this pioneering spirit, and the jauntiness of the green feather in her crushed black hat was belied by the somewhat anxious expression on the face beneath. Evacuation was all right for children, she maintained . . . she had three more in Kent, and two just outside Wisbech . . . but for herself, she'd just as soon stay at home in Rotherhithe. If the roof was seen to and the glass put back in the windows, there wasn't much the matter with the house. And they'd got a good shelter at the bottom of the garden. But her husband would have her go herself this time, and you know what men are.

By the time I left them in Port Row, we were all firm friends, and the two women had decided that they would like to remain together. On Monday evening when I called in to see them, they were on the best of terms, the cottage was 'loverly', everyone was being very kind, and could I find them a flat-iron, please? On Tuesday Mrs Waggett

called at the Manor to complain of the damp. It was a raw
and foggy day, and the Manor walls, too, were streaming.
Houses in the country were like that: they'd be all right
again in a day or two, when the weather changed, I tried to
reassure her.

On Wednesday I found Mrs Stubbings on the doorstep.

"It's the baby," she confided. "I can do with 'Ilda, but I
can't stand the baby crying. Gets on my nerves. I wish you'd
say something to 'Ilda."

On Thursday Mrs Waggett dropped in to give her candid
opinion of Sheila and Margaret. "They will tease Alan all
the time," she finished. "I suppose you couldn't find me another billet? What I'd really like would be to be in a 'ouse
with some of the people who live 'ere."

As it happened, the evacueeing friends of the landlady at
the Hart had just gone back to Yarmouth, and Mrs Grief
had murmured that she 'could do with' an evacuee or two,
so long as it wasn't school children, and they'd behave
theirselves, and she could get a look at 'em first. I suggested
that Mrs Waggett should pay her a visit. On Friday I moved
them in, and Mrs Stubbings, button-holing me in the cottage just before I drove off, thanked me effusively. With the
place to herself, she'd get on like a 'ouse-on-fire. And had I
such a thing as a clock going, as 'Ilda was the only one with
a watch?

The following afternoon I took my small travelling clock
round, to find the cottage empty, and learned from Mrs
Peel that the Stubbings had walked to Banwell station that
morning to catch the London train.

Mrs Rivers' experience was somewhat similar, except that

in her case the departing guests left earlier, and she received an indignant letter from an anonymous relative who signed himself 'A Government Inspector', announced himself 'discusted' with the whole business, threatened unmentionable vengeance for insults unspecified, and ended "Fancy sending my relation to the East Coast in any case," which was a little inaccurate, to say the least of it.

Italy invaded Greece. The blitz shifted back to East Anglia. In company with several more, the Meddenham fire brigade spent three nights (two under machine-gun fire) in putting out the blaze in some premises not far from Sedgeham. The Rivers had to evacuate the Hall in the middle of one night and stay at the Manor for the next thirty-six hours, until two unexploded bombs had been dealt with. The police rang me up at intervals to know if we had had any lately. For three hours one night the flat was filled with a deep copper glow, the result of 'enemy action' in a near-by 'East Anglian Town'. One Sunday evening, following the usual 'noises off', which are always seconded by the pheasants, the ducks on the horse-pond, and Bowes' dog, an enemy bomber raced over the Manor, a Spitfire close behind.

The Bailey family leapt into their dug-out on Pound Close, then came out again to watch. Bailey yelled frantic instructions and encouragements to the Spitfire, which had evidently lost sight of its quarry in some low-lying clouds, though we had a clear view of both. A moment later the plane jettisoned the rest of its bombs over fields in Wendeswell, and raced back again, one engine making odd sounds. Behind it the Spitfire was gaining steadily. To the rat-tat-

tat of machine-guns they vanished into a cloud-bank over Banwell Common.

"Do this sort of thing keep on much longer, I reckon we'll ha' to evacuate to London," remarked Croft, as we separated, and each went back to listen in to the six o'clock news.

Gidney, in his official capacity as an A.R.P. Warden, fitted my gasmask with a new filter, told me the newest bomb stories and taught me the correct way of handling a stirrup pump.

"And will you speak to Lambert again, miss, please? Everyone up at the cottage keeps going on about his lights showing, and he 'ont pay no regard to me."

Wheat sowing was finished. The canning company's sacks turned up at last, just as the Banwell station-master's manners were on the brink of giving way. Peachey ranged over Limekiln Breck, ploughing out likely spots. Behind him the women picked and Patfield weighed up the bags, allowing an extra four pounds to each, while I descended at intervals, brandishing a small foot-rule and measuring tops. But in spite of these precautions, as Major Rivers had foreseen, I was soon receiving a constant stream of complaints re size and excessive dirt, by telegram, telephone and letter.

Admittedly the weather was not being particularly helpful. Already this November was in a fair way to excel last as far as rainfall was concerned. But after all, dirty carrots were dirty carrots. "And as for size, you want to see some a' the ones what they ha' left in the sacks they ha' sent," remarked Patfield. "We hain't sent nothing near as bad as them."

However, four men were set to passing the carrots over the potato riddler, which knocked off most of the clinging mould, before the carrots were finally weighed and bagged up. An additional two pounds were thrown in whenever the weather was worse than usual, or a particularly 'heavy' spot was being picked over. And I bought a second foot-rule for Patfield's benefit. If the complaints from the canning company did not altogether cease, they grew more infrequent and less bitter, though a fresh epidemic broke out on the field, the women demanding an increased bonus per skep on the canning carrots as they were 'so fiddling' to pick, and the riddlers asking for something on each ton riddled, since they were not earning anything extra, and those helping to pick carrots were. I began to long for the day when the last of the small carrots should be picked.

In Port Row evacuees continued to come and go, the only family to stay on being the assorted brood of a onetime hop-picker who had been married three times and done her duty to her country and her various husbands to the tune of a baker's dozen of children, five of whom she had brought with her.

In an incredibly short time she had burnt most of the furniture for firewood, and reduced the cottage to such a state that I always took good care to remain on the street side of the door. Mrs Aldwinkle herself was incurably cheerful, not to mention entertaining. And while it was impossible to believe that any member of the family used water for any other purpose than cooking, they all looked extraordinarily healthy, and had the most enviable complexions.

Four of the tribe were under five. The fifth, eldest off-spring of her second marriage, was a sturdy youth of five foot ten, verging on sixteen years old.

"Couldn't leave 'im be'ind. Now, could I?" asked Mrs Aldwinkle when I remarked on this. "I was wondering if 'e couldn't work on the land. 'E did where 'e was evacuated last time, on a farm in Lincolnshire. 'E don't mind what 'e does so long as it ain't 'osses."

"He could try picking carrots on piecework with the women and see how he got on," I suggested.

"Seems he ain't a bad boy at all," Marsham conceded, in some surprise at the end of the week. "A pity he dussen't go nigh a horse, do he might ha' a reglar job. Still, we can do with him pulling carrots. An' do anyone wor to fall out down at the washer, he might go carrying bags out. He look strong enow."

The Fair Close oats were threshed, and ten coombs sent down to Dennington Mill for crushing. Major Rivers signed the contract disposing of the Doles carrots. The danger period for foot-and-mouth was safely negotiated, and the straw taken up in the yard. The nip in the air turned to an occasional suspicion of frost and Roger began moulding up the carrots.

"He's helping to keep down the rabbits an' all," confided Marsham gleefully the day Prisoners Close was finished. "Knocked five on the head today. That'll help the rationing."

"I suppose they're safe to eat," I ventured dubiously. "Isn't Myhill gassing round there?"

"He ain't eating 'em hisself," grinned Marsham. "Not so likely."

Limekiln Breck and the Pightles were cleared of carrots and Vincent and Leslie began ploughing them up. November's rainfall totalled nearly five-and-a-half inches. December, though it was to achieve less than half this, was to prove an irritatingly showery month. Rain caused alterations in the orders almost every day. High winds tore thatch off stack after stack, which had to be repaired by Peel and Hill, since Bushell and Meale were busy in the washer. Barretts announced that they were prepared to recommence carting carrots up to London whenever we liked. The last of the canning carrots were loaded, and I made out an estimate of the remaining crop. If we continued loading at our present rate, we could clear the lot, bar Doles, by the third week in January, I reported thankfully to Major Rivers.

"Clear the lot! You mustn't do that!" he exclaimed.

"But why not? Both London and Manchester are clamouring for them. And we're bound to get a lot of bad weather after Christmas, apart from the blitz."

"Can't be done. The market people expect you to keep on sending right into April, and that's what you've got to do. You'll have to work out a weekly average which will take you through, and only get above it if the weather lets you down the week before."

"It seems such a pity when we could get rid of them all now," I mourned. "And we shall want the women for Doles in March."

"That's the way you'll have to work it, though." The Major was adamant.

Reluctantly I revised my estimates and kept the loadings down to between twenty-five and thirty tons a week. The six acres of seed kale on the top of Mautby's Hall were inspected by another member of the seed merchant's firm, and pronounced exactly right . . . had the stalks been any larger, there would have been a danger from frost, and a likelihood of the plant's strength going into the stems rather than the seed pods, I learned. The barley stacks on Booters, Ram's Breck, and Gunspear were threshed out. The Manchester blitz burnt out a large part of Messrs Satterwaite's premises, but in answer to inquiries they wired firmly "Please continue loading." The last of the sugar beet was lifted and carted into the factory and the remainder of this season's meagre allowance of dry pulp was brought back for the cows. The Rivers children came home for the holidays, and their mother received a second letter from a prison camp near Frankfurt. Peacock wished me the compliments of the season and presented me with a bottle of elderberry wine. Messrs Root and Hopkins sent oranges from Spitalfields. It was Christmas again, the second in a Europe torn by war and deaf to messages of peace and good will towards all men. As the old year slunk off and the new year crept in, unheralded by bells, I unshuttered the flat window.

Outside snow was falling and the land lay sleeping under a thin blanket of white. Beyond the river, above Ashacre, searchlight beams pierced the darkened sky. From a long way off to the left, there came a dull muffled rumbling. Guns? Bombs? . . . thunder? Someday one would know for certain that it could only be the last.

The searchlights went out. The rumbling ceased. Only the snow fell steadily throughout the night.

≫ X ≪

"BOWES ha' started the New Year off well," announced Marsham next day. "He ha' got a housekeeper at last. Come yesterday forenoon."

He spoke as one who merely states a fact, but there was a gleam in his eye which assured me that there was more to come.

There was.

"Bin lookin' arter hisself, he hev, ever since that youngest darter a' his went into the Auxiliary Territorial Service. It's a rum un, ain't it, when the men stop at home, an' the women get into uniform? Bin took as a cook an' put through it so she's fit to cook in any hotel in the country now, Bowes was tellin' me. Do she must ha' changed a bit. I mind the time when that old dorg a' hisn fare fit ter bust out a' his coat, along a' eatin' up all the stuff what she spiled."

"Bowes doesn't look as though it did him much harm." For Bowes at sixty-three weighs fifteen stone, and is as spry and hearty as any man on the farm.

"What's the housekeeper like?" I continued idly.

"She don't look any too fierce, though I daresay she can look arter herself. Bowes, he showed her round. You know where he hang out, miss . . . the tail-end a' that row past mine, what they call the Mangel Hale. No upstairs, but a kitchen an' copperhole, a front room, an' two little totty bedrooms where there ain't room to swing a cat, all on the one level. Bowes, he's taken ter keepin' his carpenterin' an' shoe mendin' things in the room what was Dolly's, time she's gone. He wer always one for doing odd jobs an' keepin' hisself busy. That leaves but the one bedroom. 'There y'are,' he say, 'you can tek' it or leave it.' "

"And what did she say to that?"

"She's astayin'. Only don't you go an' tell t'old parson that, now. 'Tain't no consarn a' hisn. Nor yet a' mine an' yourn."

Snow continued to fall off and on for several days. There were sharp frosts at night. The land was alternately too hard and too wet to plough. In between snowstorms two more barley stacks were threshed. In the east yard at Silford Mill a heifer slipped up one night, catching a horn under the door. When Tom came to feed them in the morning, she had been trampled to death by the rest, and I had to ring up the horse slaughterer at Weddenham to fetch away the corpse . . . for which he handed me a ten shilling note. (Twenty-four hours ago she had been worth about twenty-seven pounds.) I stamped insurance cards and filled in the latest crop of forms, which included three sent from the Ministry of Labour office when Roger Marsham, George Long and Hill registered for military service, the green N.S. 100 A which I hurriedly obtained from the War Agricultural Committee when Ted Long was threatened with a medical board, and the large buff questionnaire from H.M. Inspector of Taxes, who had recently enquired if any Manual Worker on the farm had received more than £60 (if unmarried) or £100 (if married) in the six months ending October 5, 1940; and on learning that Oby fell into the former category with £67, now sought additional particulars. A stick of incendiaries was dropped on the other side of the river, between Silford Pool and Fustyweed. When Leslie was able to start ploughing on Broomhills again, he came across the burnt out remains of five more, and for quite a month no one was to have any reason to complain about Lambert's blackout. The three two-and-a-half-year-old colts, who had been kicking up their heels for the last year or two on Three Bridges or one or other of the downstream meadows, were brought up into the horseyard, and Roger Marsham and Hill began breaking them in.

They could not understand it.

All their lives they had run free and unhandled out in the open. Now they were shut up between four walls, and men thrust steal bars into their mouths, collars were put on their necks, straps and chains hung over their backs, halters confined their heads. Bewildered and unhappy, they kicked. They reared. They snapped and bit. They flung themselves to left and right. Gypsy, the chestnut filly . . . "Nappy, same as all chestnuts," Roger told me . . . had to have her legs tied, and was harnessed as she lay on the ground. For a day or two all three had to keep their harness on, and three doleful heads hung over the horseyard door, no longer seeing before them an open stretch of inviting meadow which simply asked to be galloped on, but a flint wall and not much of that, since blinkers shut out three-fourths of this strange new world.

"Must they have blinkers on?" I asked Marsham.

"Do you wor ter tek the blinkers off of any hoss on this place, he'd be over the fence afore you hed time ter tu'n round."

"But these three . . . they could be trained to do without them from the start, couldn't they?"

"We hain't got no harness without blinkers, any road. An' they're always bin bruk in that way," said Marsham stubbornly.

"Better leave them alone," the Major advised. "That's the way they're used to. And they'll have to work them."

I said no more.

Though there was still frost at night, for four days the weather was as mild and open during the day as though we were in sight of April. The biting wind dropped. The sun shone. Ploughing was started again. Kirk chain-harrowed the meadows. Several women who had begun to baulk at carrot picking started work again, and Coney Hill was cleared. Pigeons swooped down on the greens, and Frank Buxton ranged from the savoys on Horse Close and Gill Cross to the seed kale on Mautby's. Hall and back again, scaring them off. During the night following the fourth day, it began to snow again. Soon it was falling continuously. Within two or three days it was a foot deep and drifting.

There was no question now of getting the carrots away too fast. It was all we could do to send our fifteen tons a week for the next fortnight. And even this was possible only because, scenting trouble, we had managed to store well over half that amount in the carrot house before the snow began. In the barn the recently threshed barley was 'dressed'

by putting it first through the 'Boby', a machine which
winnows out a great deal of the extraneous matter, such as
'harns', dust, and too small kernels, and then through the
big 'Tasker' self-sacking dressing-machine. After which
the sacks were weighed up and piled ready for despatch by
rail or lorry the moment the forwarding order arrived from
the buyers.

A quick thaw set in, and the snow, except for odd patches
in out-of-the-way corners, vanished almost overnight. A
keen high wind made it impossible to thresh for several
days, but it soon dried the land sufficiently for Roger and
Hill to take the colts ploughing on Coney Hill. I ordered
in twenty tons of superphosphate manure for the coming
crops, and another six tons of cowmeal for the cows. Will
Long and Catton registered for military service. Croft,
Bailey and one or two more of us spent an entertaining hour
or two chasing parachute flares in an atmosphere heavy
with the smell of burning resin. But next morning I had to
surrender to the Major, in his official capacity as a Home
Guard, the one which came down on the glebe. Roger Marsham, Cecil Peachey, Kirk, Meale, and Frank Buxton went
down with flu. I lent out, as usual, my weekly review to
Bailey, the third cowman, whose father had once spoken on
the same platform as Keir Hardie, and who was himself,
in spite of being the possessor of a mind which worked at
the pace of Oby's, an ardent politician with a surprisingly
wide and accurate knowledge of current affairs.

Generally I left the review on the shelf outside the back
door with the milk jug, the moment I had finished with it.
On this occasion, running into Croft, who had just brought

some potatoes into the scullery, I suggested that he should give it to Bailey, since the two lived side by side; then the latter could begin reading it in his dinner hour if he liked, instead of waiting until the evening.

"Read? What, Billy? It's his missus as is the scholar an' reads out them papers a' yourn." Croft, a strong conservative, though he has never had anything worth mentioning to conserve, is always properly scornful of anyone who holds views opposed to his own. "Billy, he can't read at all. Do that ain't more than a word or two he's taught hisself to spell out at a time."

In this year of grace, in Mother England! I could not believe it. But it is true.

"You see, I was allus a bit slow like," Bailey told me. "Tuppence short in the shilling, me feyther used to say. Though I don't think tha's right. It wor just that I worn't as quick as the rest on 'em. An' when I went to school, that ud be about twenty-five year ago, the teachers, they'd got about sixty on us in a class, an' they hadn't no time for the ones same as me. We hed ter shift fer ourselves. Now, my missus, she allus wor a sharp un. Worked on the land las' war an' got a month's scholarship to one a' them there agricultural colleges, an' all. She still hev all her books. But I'm agettin' on. I kin do a bit a' the papers, do I take me time over it. But it's a slow job."

Oats must be in by the middle of February at the latest, announced Major Rivers. So we threshed out the stack on Easter field for seed to sow at once. But as it turned out, March was to come in before the land was fit to drill.

The newspapers and the B.B.C. were full of the new ra-

tioning scheme for livestock, and since we had received no
official intimation of it, I borrowed and studied a pamphlet
on the subject from George Long, who kept a couple of
hundred hens, and had. I ordered new piston rings for the
engine in the barn and for the carrot washer engine, and
let Bushell, Mark and Paynton's engineer loose on them.
And I did my best to deal faithfully with the latest official
request for information — this time a list of every item of
live and dead-stock on the farm.

The first was child's play, but the second was to take up
all my spare time for nearly a week. With Marsham I
toured every cartshed, barn, and outhouse on the place, noting down everything we discovered, some finds being a surprise to Marsham himself. For the first time I learned that
harrows are of countless sorts, wood and iron framed,
'duckfoot' and 'disc', spring and chain, single or in 'gangs',
and made up of various numbers of 'baulks', the sections
which regulate their width.

"Then there's the question of spare parts, especially for
the tractors," the Major remarked, when at last the list was
done and a copy filed. "What have you got in the store?"

The 'store' is an elastic term which covers the late outdoor
office, the one time electric power house, the fitter's and
carpenter's shops, the miscellaneous assortment of ex-hen
houses and such adjoining the corn barn and known collectively as the black 'huts'. The job of listing their contents had been hanging over my head for months, but had
never been completed owing to the fact that Mark's spare
time and mine had not yet coincided for long enough to
get down to it. Marsham had done his best to teach me to

recognise such things as 'eccentrics' and coulters for drills, 'keeps' for the power-binder, 'wings' and 'points' and ribs for the Cambridge roll, 'fingers' for the grass-cutters and sundry other mysterious objects. But the tractor parts were nearly as much of a closed book to him as they were to me.

"Half the tractors of the country will be laid up before this war's over. And all for the want of some silly little spare part," the Major went on.

"I met a man the other day who'd just had to buy a new Fordson because he'd broken the crankshaft on his old one. He could get a new tractor within a fortnight, but they wouldn't promise him a crankshaft for a couple of months. Spare parts will always come in handy. You'd better talk to the men as you go round, and ask them what they think they can do with during the next eighteen months or two years. Then order it. You won't be able to get half you order, but you'll get something."

The weather continued to swing from one extreme to another. One day was bitter and sunless, snow lay a foot deep. The next it rained, and the wind rose to gale height. The day after that was like late spring, and most of the snow melted. Next morning the post included our 'rations for stock' papers, with the polite intimation that the Manor had been classed as a 'surplus' farm, and until we had threshed out and sold all our corn and could produce the sales tickets, we were not entitled to any food for our stock at all.

"It's my belief that all these rationing schemes are worked out in Bedlam by the oldest inhabitants," said Major Rivers. "Give me that telephone. What's the number of the War Agricultural Executive Committee?"

The war in Greece went on, with the Greeks taking the offensive. In Libya, Benghazi fell into our hands. The Royal Navy bombarded Genoa and Ostend. Britain broke off diplomatic relations with Roumania. There was a crisis in Bulgaria, already practically occupied by German troops. The R.A.F. continued their raids on Western Germany and invasion ports in France. British paratroops landed in Calabria.

Above the familiar six A.M. rattle of milk churns the following Sunday morning I heard a plane passing low over the Manor with one of its engines missing. A few seconds later though no one in Silford, but Gidney, was to know it for half a day, and Gidney not for another hour, the crew had taken to their parachutes, three coming down at intervals between Dennington and Norwich, and the first to 'bail out' landing on Palegate.

"You can see his heel-marks an' the place where he lay afore he come to hisself, right in the middle a' the wheat. I ha' bin an' hed a look," Gidney related next morning. "Me an' my missus, we wor abed an' asleep, when there come such a knockin' an' a hollerin' at the door. I looked out a' the winder, an' see someone on the doorstep, but 't worn't light enough ter see who 't wor.

" 'I'm an English airman,' he said, drectly he see me.

" 'Are you sure you're English?' I want ter know. Do I was off down stairs to get hold on my .410 what t'old Guv'-nor give me to scare rabbits with. My missus, she never see no fear, she'd got downstairs an' the door open by that time, so he'd ha' shot her fust. An' in he come, plastered with mud. Seems he lay on Palegate nigh an hour. Then he struck towards the Silford Hill buildings, taking 'em to be

a house, missed the roadway when he found out they
worn't, an' dragged hisself right across that sticky old muck
of an Easter field to mine. He et up six rounds a' bread-an-
butter, two eggs an' put down five cups a' tea afore he made
off to Dennington to telephone where he was. Only a young
chap. 'Bout twenty-one. His home's in London.

" 'Now don't you go tellin' me anything as I didn't ought
to know,' I say to him. 'But I'm an A.R.P. warden. An'
I'd like to know what you've bin up to, afore you come
down here.' Bin to Germany with four bombs each weighing a ton, so he said. Had engine trouble there, so got shut
a' the bombs an' was nearly home again when they caught
on fire. I wonder where the next'll drop? Things allus go in
threes."

Two days later Gidney, who must be adventure-prone,
was carting in a load of waste carrots for the cows from
Prisoners Close, when a Whitley bomber roared past his ear.
When he managed to pull up his runaway horse and look
round, it was sprawling across the fence between the
Pightles and the lower glebe, where it was to lie for the next
three weeks or so like some enormous deep sea monster
beached by the tide.

No one was hurt. The crew and some military passengers
picked up in Malta a few hours before, stayed to breakfast
and presented me with treasure trove in the shape of a pot of
marmalade and half a dozen oranges and lemons. Whilst
they awaited a rescue party from the nearest aerodrome,
they allowed me to crawl round inside the plane . . . a con-
cession which was refused all subsequent comers, including
Major and Mrs Rivers, as by this time an R.A.F. tender had

fetched away the crew, and a young and highly-conscien-
tious military guard was in possession. The third of the series, in spite of Gidney's prophecies, has yet to happen.

With the Major I got out the first tentative cropping list,
and then worked out the amount of seed required and the
artificial manure needed for the various crops. I button-
holed in turn every man on the farm who touched machin-
ery, and tried to prise out of him a rough idea of what parts
of his machines were likely to wear out during the next
eighteen months or so. Hardly one would commit himself
to anything like a concrete statement.

"That all depend," they reiterated.

"But you must have some idea!"

If they had, they generally managed to keep it to themselves in the same maddening way in which, whenever I
tried to find out how long any particular job would take, I
would usually get "That'll be a bit a' time yet," or "That 'ont
be as long as all that," with no indication as to whether ten
minutes, ten hours or ten days would fill the bill. But finally
I achieved a list of sorts and forwarded the result to Cawston's at Meddenham and Weldon's at Sedgeham and the
'store' slowly filled.

We began to thresh the Upper Common barley stack, but
had to stop with it half done because of snow. Just as I was
telling the men to leave off, a low flying raider skimmed
out of the mist, machine-gunning as it flew, and we all fell
flat, the quickest to achieve the horizontal being Will Long,
who had just come back from a Home Guard course which
had included a demonstration of dive-bombing.

The latest official ruling for marketing of stock decreed

that all beasts must be declared beforehand to the auction-
eers, when the latter would fix the date for their disposal,
which must be kept without fail or heavy penalties would
result. . . . And on declaring six of the Silford Hill fat cattle and two culled cows, we were graciously accorded permission to take them to Meddenham market in twenty-one
days' time.

"We've no business to be growing wheat after wheat on
that far end of the Upper Prisoners," said Major Rivers,
while he revised the cropping list. "But it's only this once.
Most of the Prisoners was peas last time. And the land's
been well manured. Still, we mustn't do it again. As a rule,
we more or less follow out the rotation of crops system.
Wheat, then roots, then barley, with small-seeds drilled in
for hay the following year. Then start all over again with
wheat. But you can't keep strictly to that, especially with
market-garden stuff. So roughly we grow barley, carrots,
barley, carrots, on the light land; and wheat, savoys, barley,
on the heavy. A root crop gives you a chance to give the land
a good cleaning. That's why we've put those two biggest
glebe fields down for sugar beet and carrots. And a nice job
you'll have with them. A lot of the foul grass has rotted in,
but there's plenty left. It'll be a race between the weeds and
the crop all the time, with the weeds two fences ahead from
the start."

Vincent ploughed and then immediately rolled the small-
est glebe field, which was destined for oats. Had we intended to sow barley there, he could have left it on the
plough for a bit, Major Rivers informed me. "But wheat
and oats like firm well-rolled ground. And an olland

doesn't break up like ordinary land. If you left it and it rained, you'd have a job to get it rolled later on."

The Army called to enquire if it could fight a practice battle over the farm in three days' time, and promised to avoid the wheat and carrots. But it snowed again, so they called the battle off. The infants' mistress at Ashacre school, driven to desperation, sent one of the young Aldwinkles back to his mother with a request that he should be bathed, and received in reply a note which ran: "Dear Miss Blyth, Terence smells just like his father and I've slept with him for thirteen years. You don't know what love is."

Peel and Kettering took the engine and chaff cutter off to Ashacre Hall to cut up a stack of oat straw.

That evening the Major rang up.

"I shall have to borrow two men off you tomorrow. Any two will do."

I caught Marsham in the yard just as he was going home. "Which two shall we send?"

"Hill and Catton live nearest that way. Peacock's putting in some overtime tonight on that wagon shaft. He kin warn 'em when he go home. An' Cecil Peachey could take Catton's place in the washer jest fer the once. I kin ketch him when I get home meself."

As I was bolting my breakfast next morning, Marsham appeared at the door.

"That fule of a Catton!" he began. "There he is, jamming about, round by the motor-house."

"Catton? But what's he doing there? Why isn't he at the Hall?"

"I ha' hed ter send Kirk instead. You'd better go an' hev'

a word with Catton. He reckon he want ter see the Guv'nor an' hev his money an' go. Do he see the Guv'nor in the mind he's in now, he'll go all right. The Guv'nor'll see to that. An' he can't afford to throw hisself about like that, with a wife an' family."

"But what started all this? He's been to the Hall plenty of times before."

"Seems as though Peachey told him in the 'Ringers' the night afore last as he'd have him out a' the washer this week, and Cecil in. You know how touchy Catton is, along a' him bein' deaf. An' Peachey, do he's hed one or two, fares as though he must go agarawatin' someone. Tha's eggs an' milk to him."

It took me nearly half-an-hour to convince Catton that coincidence, not Peachey, had resulted in Cecil's being in the washer that morning.

"But how did Peachey *know?*" he kept on doggedly repeating.

"He didn't know," I reiterated exasperatedly. "I didn't know myself until last night. And you'll be back in the washer tomorrow."

"But, Peachey said he'd get me out and Cecil in an' he's done it."

"No, he hasn't. But if you're mug enough to sack yourself, Cecil will stay in the washer," I pointed out for the sixth time, keeping one eye on the lane for the Vauxhall.

In the end I won by eighty seconds. For Catton had barely disappeared from sight on his way to join the carrot pickers for the rest of the day, when the Major turned in off the Ashacre road.

March came in with a wind which cut like a knife. I had

to ring up the horse-slaughterer again, this time to fetch away Short, now nearly nineteen years old, and at last past work.

"Ont be long afore he ha' ter call agin," prophesied Marsham gloomily when, having witnessed the firing of the fatal shot, and Short's final departure on an open lorry, he handed me over the pound note which completed the transaction. "There's Gilbert, an' Captain. They don' neither on 'em get any younger. An' that leg a' Brisk's don't fare to get a lot better. You might write a pus'card to old Jeffries, an' ask him to come an' hev another look at him. Still, them t'ree youngsters aren't comin' along too bad at all, by the look a' things. Tha's more than I kin say fer them mucky young toads a' boys. Barrin' young Evan. I can't say as I ha' got a lot a' fault to find wi' him. But that Sam evacuee, he ain't tu'ned out a lot a' cop. Late agen these last t'ree mornin's. An' as fer that there Frank Buxton, I don't know what ter say about him at all. Do you put the two together, there ain't a hen's noseful a' work done between the two. But Frank's wuss than ever on his own. I wish you'd hev another word with Mark. Do that hed a' bin anyone's else's lad, I'd ha' kicked his backside an' him off the place long since."

I had already had more than one word with Mark. I had also said a good deal to Frank on sundry occasions when I had come on him smoking in a pithole instead of picking waste carrots, or singing happily in a ditch when he was supposed to be scaring pigeons. As a last resort I had set the Major on him, but even the effect of this last had worn off within twenty-four hours.

"He really must alter a bit, or I shall have to speak to the

Major again. And you know what that will mean," I hinted delicately to Mark, having run him to earth in the fitter's shop.

"Blast the boy! He's like an old dickey. The minute you leave off thwacking him, he stop makin' haste," returned Mark dejectedly. "I don't know what it is. Seems like he's a different breed from the rest on us. He never ail anything. He just don't fare to trouble. Don't matter what you say or do to him, it's all the same. Put a new suit on him th' other Sunday. Out he went, an' what must he do but go straight off an' jump over one a' them there tar barrels down the Grove Lane. I was sittin' indoors, readin' the Sunday paper, when I hear someone sniffin' on t'other side a' the door.

"'What's that?' I called out.

"There didn't come no answer, but it went on. So I went to see an' there stood Frank.

"'Oh, Dad, I aren't half in a mess,' he say. He wor an' all. Cor! I whatcher call clawed hold a' him, an' socked in ter him lik flyin'. He'd ter go ter bed fer a fortnight drectly he'd hed his tea. But that didn't fare ter trouble him a mite. Not him. Went off every time lookin' as though he'd lost a tanner an' picked up half a thick un. I'll drop on to him again, miss, though do you'll give him another chance."

"The trouble is," said Patfield thoughtfully, as we worked out a scheme for retrieving carrots on piecework which would at least ensure that any idling resulted in a monetary loss to the culprit instead of the firm, "Mark is always on to Frank about something. So the boy's got hardened. I ha' only got to shout at mine once, course it ain't often I say anything to 'em."

As far as Sam Aldwinkle was concerned, the problem was temporarily shelved by his cutting the top nearly off his thumb instead of a kale stalk next day. On piecework Frank worked so furiously for a week that he earned nearly double his ordinary wage. When there were no waste carrots, we made a point of putting him to work with someone else. He drove the one-horse roll in front or behind the drill. ("Roger 'ull see as he keep on the move," said Marsham grimly.) Sometimes he took a turn at grading carrots in the washer. Making up the paybook it dawned on me that the net result of Frank's being a thorough nuisance to everyone, was that he ended up by being alloted most of the jobs to which small bonuses clung. Whereas Evan Gidney's sole reward for being reliable and hard working was that he spent practically all his time cutting and carting in kale for the cows, to which no extras were attached.

"That ont do," Marsham agreed with me. We spoke to the Major and Evan's wages were raised.

I sent in some more corn sales tickets to the Feeding Stuffs Officer, and asked wistfully when we were likely to receive food coupons for our stock. (Already the B.B.C. was announcing that the unit would shortly be reduced from one cwt. to half, and so far we had not received a single coupon. Had we not had several tons of cowmeal in stock before the rationing order came into force, the cows would have starved long ago.) I filled in a form demanding sugar for the two hives of bees in the orchard. I filled in another, as comprehensive as a Ministry of Agriculture return, on behalf of Harold Peachey, who had written to ask if we would apply for him to be granted twenty-eight days' agricultural leave. The disposal firm which had been clearing away the Whitley bomber, requested my signature to yet another, to say that the field had been left in an irreproachable condition. Roger Marsham started work once more and drilled oats on the glebe, with the filly colt, Gypsy, harnessed in the drill. The following morning I was wakened just before six by all three colts racing down the drive past my bedroom window. By the time I had thrown on a dressing-gown, they were trotting solemnly up and down the tennis lawn, their hooves tearing up the turf in fresh places at every step.

As I did my best to shoo them off, Bailey suddenly appeared from the direction of the kitchen garden, remarking that he thowt as somethin' wor up; and together we managed to drive them through into the orchard, and then open the gate on to the Walnut Meadow, where I left them for Bailey to get back into the yard. Neither Marsham nor Peachey, one or both of whom were responsible for this state of affairs, came to the rescue. One of them, presumably, had stayed behind to prevent the rest of the horses from following the colts. But where was the other? I should have something to say when Marsham appeared under my window just before seven.

It was nearly half past eight before I saw him, however, when he had to own up to another 'turn'.

"They mostly happen of a Sunday," he told me aggrievedly. "But I wor caught out properly this mornin'! I could see them colts from outside my back door, but I couldn't do nawthin'."

"I shall ring up the doctor myself soon," I threatened.

"What, ha' you bin ailin'?" asked Marsham innocently.

"You want ter tek care a' yourself, you know, miss. Don't
you go an' get laid up, whatever you do. I ha' told Flip to
get a couple a' iron hurdles into the yard, agin the riding-
stable, to stop the colts in future. An' there's ten shillun'
you want ter put Peachey down for a' Friday. Beauty hed
her foal at fower o'clock this mornin'. Do you come an'
hev a look at him."

Overnight the bitter weather of the last ten days changed
again. The wind dropped. The sun came out. Birds sang
and began to build their nests in fence and bush. Roger
drilled barley on Coney Hill, the upper glebe, Clam Breck
and the Well Piece, with Frank scuttling ahead of him with
a one-horse roll.

"Must roll in front for barley," Marsham told me. "That
make all the difference to the drillin' an' the sample. Do
you get barley fleet in fleet an' deep, deep an' fleet, that ain't
no good at all."

The last of the glebe was ploughed, the sandy subsoil
thrown up in patches by the 'digger', making the field next
the Pightles look as though some sportive giantess had
scorched it with an enormous flat-iron. Feeding stuff coupons arrived at long last, allotting us less than three quarters of our normal requirements, Beauty and her son were
turned out in the small grass enclosure on the Barn Breck,
adjoining the Church Lane, and after several abortive at-
tempts, the foal suddenly discovered that he could run.
Thereafter he tore round and round until he was exhausted,
while Beauty fondly looked on. We started lifting carrots
again on the Lower Prisoners and Doles, turn and turn

about. The day the first five tons were loaded from the latter, Mr Armstrong, the new buyer, rang me up at ten P.M., to know if we would wash them in the future.

"No!" exclaimed the Major emphatically, when I rang him in the morning. "They were sold dirty and dirty they go."

"He's offered to pay for the washing."

"How much?" inquired the Major suspiciously.

"Ten shillings a ton."

"That'll hardly cover the men's time. Then there's a difference in the weight. A ton of dirty carrots won't make a ton of clean ones. And you haven't got the men to spare."

Two days later Mr Armstrong called to take another look at Doles, and again broached the subject of washing.

"As you know, we're loading to the order of the Ministry of Food. And they've had a lot of badly frosted dirty ones in."

"Not ours," I said firmly.

"No, not yours," he admitted reluctantly. "But the result was the same. The Ministry is practically refusing to accept any dirty carrots for the next two or three weeks. Now if they were washed, it would be a different matter," he added persuasively. "You want to get the field clear, don't you?"

"The contract stipulates regular loadings every week," the Major, who had just joined us, reminded him blandly. "And if we could find the time to wash you some, it'll cost you a pound a ton," which effectively closed the question for the time being.

That week we loaded seventy tons of dirty carrots from

Doles, and early on the following Monday morning Mr Armstrong rang up to say ungraciously that we could rail sixty tons during the coming week. I worked out the London and Manchester lists to fit in with this, arranged for trucks and lorries, and settled with Patfield which field should be worked on which day. At eleven o'clock that night, Mr Armstrong rang up to say we must not load a ton over thirty-five unless we washed some. So next morning I made fresh plans, and arranged to load more to London and Manchester, who had been demanding heavier loadings for days. As washing was finishing that afternoon the telephone rang time after time.

"Are you Ashacre 12? A telegram from Spitalfields . . . Boro . . . Covent Garden . . . Manchester. . . ." They were all the same. "Stop loading carrots until advised." I rang up Barrett's to make fresh arrangements for lorries and we began cutting savoys on the main-road end of Banwell Close.

R. A. Leeds, calling in the day after we began, suggested that we should pack some of them in half-bags, like the carrots, instead of the larger cabbage bags. "They should sell well that way," said he.

Will Long, in charge of the savoy gang, told me stories of the fabulous prices cabbages were making in London.

"I was talking to a man what's just come down from there. Reckoned they hardly ever see anything green. And payin' ten pence a pound for stuff not half as good as we ha' got here. Savoys ought to fetch a good price t' year."

The cabbage bags hold roughly about seventy pounds; the half-bags forty. Leeds returned one and six pence per half-bag for his first fifty, and deducted four pence ha'-penny per half-bag commission. Root and Hopkins made one and three pence of theirs, and took off five pence. Full bags fetched from three to four shillings less six pence and eight pence commission. Major Rivers began to talk about ploughing them in. "At least they're worth something as manure," he said grimly.

It began to rain again. One night eighty-six points fell. Then the weather cleared again. I stamped up and issued unemployment cards for the men to register for the extra cheese ration promised to agricultural labourers. Invasion was in the news again, and Isbell and Sam, whose thumb had now healed, spent two days filling up new sand bags for the Home Guard. Bombs fell on a small 'East Anglian market town' nearby, demolishing the Salvation Army Hall, but missing the adjoining licenced premises, the land-lord of which went round to commiserate with the remark, "Seems as how all your prayin' didn't do you no good arter all, together," so Peacock gleefully reported to me. I re-ceived an official notice headed "Military Training" which bade me please note that troops would be training in my neighbourhood on the fourth of April, and would I there-fore make what arrangements I considered necessary for the safeguarding of my live-stock and property. Should any damage be done, full particulars must at once be sent to the undersigned, who would inspect and deal with the question of compensation, should it arise.

I caught both Sam and Frank having a free fight with waste carrots on the Lower Prisoners, instead of gathering up a load for the cows. When Marsham went down to fetch

Frank off for another job the morning after, both were missing. They were still missing when he went down an hour later. Major Rivers, meeting Sam coming out of the Hart as he was driving past, sacked him right away. Frank, eventually discovered asleep at the New Buildings, was given a final fortnight on probation. Mr Armstrong called again, and once more raised the question of washing carrots.

"I believe you'll have to, in the end," the Major said, when he had gone. "Or you'll never get rid of them. There seems to have been a thorough mess-up, up in London. The Ministry have been trying to store them, so Leeds was telling me. And of course they began to rot. Now they're letting them all on to the market again. Still, we'll hold Armstrong off as long as we can. You really have not got the men to spare."

Roger drilled barley on the Pightles and ploughed the furrows back on the glebe. On Gill Cross Vincent began ploughing in the rest of the savoys. March went out as it had come in, sullen and sunless, with a cold keen wind blowing in from the sea.

❧ XI ❦

ON my way home from Horse Close one sleety April day, I went by way of Silford Hill to make sure that old Tom knew that he must keep all the cattle shut up in the yard for the next three days, in case they should get mixed up in the 'battle'. I found him wheeling his bicycle away down the Sandy Loke.

"You'll ha' ter send someone else to see arter the cattle. I'm leaving off," he answered gloomily, when I asked where he was going. "I aren't feelin' any too cracky. What's more, I kin smell the earth."

"Good gracious! You don't feel as bad as all that, do you?" I asked anxiously.

Tom suffers from intermittent attacks of rheumatism and had, I knew, been in the throes of one for the past week. But while he looked pinched and peaked, so did everyone else, for that matter. Working out in an east wind is hardly to be recommended as a beauty treatment. But he seemed far from dying.

"Well, tha's earthritis I ha' got, ain't it?" demanded Tom. "I kin smell the earth, I tell yer. An' I'm off home until that stop," and he wheeled his bicycle off.

April had come in escorted, not by the soft breezes and

flower-enticing showers of which the poets sing, but with
an arctic east wind which was, as Marsham said, "enough
to tek yer hed orf." It painted noses purple, reddened ears,
set eyes watering and tempers on edge. Master Pepys tore
round like a mad thing, in and out the house, up the front
stairs, skidding with all claws out down the linoleum covered passage to the flat, then whisking down the back-stairs
into the kitchen and out again.

One morning R. A. Leeds called in person to say that the
savoy trade, which they had pronounced brisk the day before, had collapsed again. Hundreds of bags were rotting in
the markets. As for carrots . . . half the little Covent Garden fruit dealers, not being able to get hold of any fruit, had
followed the Government lead, and stored carrots, then
held them for a rise in price. "You can guess what's happened to them!" he finished.

I could. Even the Doles carrots were beginning to rot
with the continual frosts and bad weather. Frank Buxton
had a full-time job picking out rotting and frosted carrots as
they came out of the washer. The market returns were full
of complaints and deductions for spoilt carrots. We had
ceased to load to Manchester altogether, because trucks
were delayed so often on rail that half their contents were
frozen and useless on arrival.

"We are loading our last to London this week. And we
shall finish loading to the Ministry of Food early the week
after next," I told him thankfully.

When he left I rang up the Major about tomorrow's
savoys.

"You'll have to stop cutting. And Vincent must plough

the rest of them in," he ordered. "Is Marsham anywhere about? I want a word with him."

"Yes. He's just coming to do the labour books." I held out the receiver. "Here you are. The Guv'nor wants to speak to you."

Marsham backed hastily away.

"I hain't never spok' down one a' them there things," he said nervously.

"It won't bite you."

Very reluctantly he came nearer and took the receiver from my hand, holding it gingerly up to his ear. I heard the Major ask a question. Taking a deep breath Marsham replied with a shout that made me jump nearly out of my chair. The conversation proceeded.

"Why, there ain't nothin' to it, is there?" cried Marsham triumphantly a moment later, when he handed the receiver back to me.

So far we had seen nothing of the threatened manoeuvres.

Shortly after Marsham's departure, however . . . so soon, in fact, that I thought he was back again . . . there was a knock at the front door. When no one appeared in answer to my call, "Come in, Marsham," I went into the hall, to find a short stocky soldier by the open door.

"I'm looking for somewhere for me orficer to sleep, ma'am. Could he come in here?" he began. "We'll be all right in your buildings. But I must find somewhere else for me orficer. He mustn't be in with us."

"There's the kitchen," I suggested dubiously. "Come in and have a look at it. Most of the house is shut up."

"It'll do me orficer a treat," he informed me a moment later.

"Right you are. I leave the front door unlocked. I shall be upstairs if there's anything you want." I turned to go back to the office, but the army lingered.

"I suppose you haven't got such a thing as a cup of tea going?" he asked wistfully.

Before the kettle boiled six more men had appeared at the front door and eleven at the back. Whilst the second kettle was boiling and I hunted out the mugs left over from Silford's Jubilee celebrations, the front door bell rang again. And I found 'me orficer', who looked as though he ought to have been in the fifth form instead of the army, on the doorstep. Had I, he enquired diffidently, seen any of his men?

Quite a few of them, I told him, and led the way back to the kitchen, where practically the whole platoon had now assembled or overflowed into the scullery.

Outside the rain was falling steadily. Over more tea and the last of my cake and biscuits, we discussed the question of sleeping quarters for everyone. There were the cart sheds, the riding stable, part of the corn barn, and sundry other corners where they would be dry if nothing else. "But do be careful with your cigarette ends and matches," I implored them.

"The Army always pays for any damage it does!" exclaimed 'me orficer', immensely shocked.

"And a lot of good that will be if you burn half the buildings and the implements. Paying for them won't replace them," I rejoined tartly.

"Might they borrow some wire-netting and old hurdles to cover in a trench or two?" 'me orficer' enquired, when I had shown him where the various sleeping apartments were to be found.

"Where are you digging trenches?" I asked suspiciously, as I shooed him away from the sets of harrows in the lean-to in the stack yard and pointed out the hurdles. "Not in the carrots or our newly sown corn, I hope."

"That shouldn't be necessary," he assured me. "If you'll just tell me where they are . . ." which I promptly did.

Throughout the night lorries rumbled through the yard, men shouted orders, motorcycles roared down the drive, rain still fell. Next morning I looked out of the window to see a gun and an observation post on the tennis lawn. Croft was already escorting conducted tours round the premises. At half past seven the tea queue began. By nine o'clock it was hot water for washing, and could I sell them milk and eggs. On the way down to the carrot washer I had to pass through the three dannert barriers. Bren carriers lined the roads. At intervals men crouched against the hedges under huge nets festooned with brown and green rags. Field telephone wires branched off the turnpike to run down the Ashacre road to the gig-house, and then half way up the Church Lane to where H.Q. was located in a pig shed. More wire twisted across the Barn Breck from the bottom of the Sandy Loke to the Church Lane. The main road end of the latter was choked with it. There were trenches on the cleared part of Doles, on Barn Breck and Pound Close. With much bitterness I pointed out to 'me orficer' that, in spite of all his promises, two large holes had been dug in the newly sown barley on Coney Hill.

"Not holes," he protested. "You should never talk to the Army about holes. They're weapon-pits."

"Weapon-pits or holes, they were still in the middle of our barley," I insisted peevishly.

In the afternoon I had to go to Dennington Mill and then on to Meddenham for tractor parts, so I promised to report any 'crash helmet' troop movements that I might discover. Just before reaching Meddenham I ran into a whole detachment of the 'enemy'. On the way out I passed two of their tanks. Arriving at 'Nowhere', still in the hands of the 'tin hats', and bursting with information, I was cut short by a bantam sergeant who sternly demanded my identity card: the first time I had ever been asked to produce it.

"Your driving licence, then," he said sourly, when I had to admit that I never carried it. The latter, too, was in the flat, of course.

The sergeant looked farther and farther down his nose.

"Of course, I'm not the civil police . . ." he began.

"If you shoot me at dawn, not another cup of tea will any of you get. And, anyway, I have been spying for you," I said indignantly.

Finally I was allowed to proceed to the Manor under escort, and reached it just in time to witness the capture of an 'enemy' lorry, loaded with machine guns, and three escorting motor-cyclists, who had lost their way and come into the Manor yard by mistake.

At six P.M., the original contingent marched off and a new and larger detachment marched in. Old Tom, who keeps a few fowls in discarded motor-van bodies on the edge of Silford Heath, found the troops curled up in his hen roosts and his Rhode Island Reds perching on the gorse bushes. The carrot house held two platoons. The Manor was commandeered for G.H.Q., which begged successively for the bathroom, aspirin, and the use of the telephone. On the kitchen front I was forced to point out that there was such a thing as rationing, and future comers must bring their own tea leaves. Once again the night resounded with the grinding of gears, the flying departures and returns of despatched riders. In the morning the yard was so full of men forming fours that I could hardly get through with Amy. On my way up to Gill Cross to look at Vincent, I passed over seventy stationary lorries, a man asleep over every steering-wheel. They looked as though they were there for the duration. But by three o'clock in the afternoon only the coils of dannert, the telephone wires, patches of oil and ruts in the roadways, empty cigarette packets and weapon-pits remained.

"The Army is coming back to clear up, I gather. But there's no knowing when they'll do it. Who can you send to fill in those weapon-pits on Coney Hill, and put in some barley on top of them in the morning?" I asked Marsham, as we finished the labour book.

"Will and Catton ull do fer that. There was somethin' else I wanted to talk about, I know." Marsham scratched his ear.

"You'll remember it when you get home, I daresay. It'll do in the morning."

"Can't call it to mind, whatever 't wor," said Marsham reluctantly, after a pause. "But I know there wor somethin'." He went slowly off down the hall. Then suddenly came marching back.

"I ha' got it. What d'ye think a' this here Benghazi business, miss?"

The news that morning had been full of the British evacuation of that recently captured Libyan port. When we took it, it had apparently been of first class importance. Now, according to the B.B.C. and the newspapers, it had none at all. Like every one else, I didn't know what to think. Nor did Marsham, though this did not prevent us airing our views.

"We want ter look out. Before we know where we are, we'll be retreating again. Like that there Dunkirk business. I don't like the look on it at all," Marsham summed up. "Still, that 'ont do no good me stayin' here. Do my missus 'll start mobbin' if I'm late in fer me tea. So I'll be gettin' along. Goodnight, miss."

The war news continued to be as depressing as the perpetual east wind. The Germans invaded Jugoslavia and Greece. Though British troops reached Addis Ababa, elsewhere in Africa they were falling back. On the other side of the Mediterranean, Salonika fell into enemy hands. In England Bristol and Coventry were badly blitzed, the latter two nights in succession.

"I ha' got three sisters live in Coventry," remarked Croft thoughtfully the morning after the second. "I wonder how they're gettin' on."

"Haven't you heard anything? Have you written to ask if they are safe?"

"Don't know the address a' any on 'em," replied Croft cheerfully. "To tell you the truth, I'd pass Maggie in the street without knowin' her. I hain't caught sight on her for twenty year. But I don't suppose they're any on 'em took much harm. Do I shall soon hear. Bad news always get round." And he went off whistling.

I wrote two letters to the War Agricultural Committee, one enclosing a copy of our March milk yield with an application for extra feeding-stuffs coupons, the other for a permit for five cwt. of barbed wire, the least we could manage with for the glebe. In due course, the latter arrived made out for a quarter of that amount, and even that took weeks to come by, for no ironmonger could get hold of it. Meanwhile coils of dannert hung . . . and still hang . . . rusting on half the Manor fences.

I acquired . . . or perhaps I should say 'knocked off', . . . a steel helmet, and 'knocking off' some implement paint from the carpenter's shop as well, I camouflaged it with an all-over design of red and blue flowers, with two carrots rampant on one side for my 'divisional sign'. The Army reappeared to fill in the weapon-pits and ask for more tea. Roger finished sowing barley on the Hart Piece the afternoon before the Good Friday holiday, and both started and finished the seven acres on Leasepit Breck on Saturday morning. On Saturday night Lambert and Peachey celebrated the holiday by coming to blows in the Ashacre 'Ringers' at closing time; and continued hostilities with such thoroughness in their joint back yard when they got home that Marsham had to get up and separate them. "A nice old how do yer do," he told me the following afternoon. "Wok me up at half arter one. Blood all over my winder-sill, an' they don't half look a pair a' beauties this morning, both on 'em. Jest you tek a look at 'em next time you see 'em, miss."

For the second Easter Monday in succession Blue Peter wandered unheeded over the Three Bridge Meadow, instead of tearing over the steeple-chase course at Fakenham cheered by a roaring crowd. While Marsham spent another fruitless night sitting up with Kitty, whose foal, due to arrive more than a week ago now, was obstinately remaining unborn.

"Pity we can't fix up some a' them there telephone wires the Army ain't fetched away yet, so's the old girl could ring up an' let me know," he said next morning. "Tha's the fourth night I ha' set up fer nawthin'. That'll tu'n out ter be a colt right enow, comin' late like this. Do you see."

Sure enough, it was. Turning out drawers in my bedroom three evenings later, well after leaving-off time, I heard sundry unusual noises coming from the direction of the horseyard. I leaned out of the window.

"Is that you, Marsham? What is it? A girl or a boy?"

"Tha's a boy all right. A right nice little feller he is, an' all. Beat Beauty's into a cocked-hat. Come you an' hev a look."

In a box adjoining the horseyard Kitty stood gazing proudly down on her coal black son. Marsham, no less proud, gazed too.

"Ha' you told the Guv'nor?" he greeted me.

"He's Home Guarding tonight," I prevaricated, knowing the Major's opinion of any but absolutely unavoidable telephone calls.

"Ar, well. He'll see him in the morning," said Marsham, after a pause. "Now if you ha' got such a thing as a drop a' hot water in the house, I'll mix the old girl up a bran mash." Reluctantly he tore himself away.

I wrote two more letters to the War Agricultural Committee, the first to the Pests Officer, for a half price permit for Cyanogas for rabbits, the second for another form N.S. 100.A. (For Oby, this time.) Kirk began sowing a mixture of muriate of potash and sulphate of ammonia on Three Bridges, in preparation for sugar beet. Leslie began to plough up the cleared part of Doles. Frank Buxton picked out kale stalks on Pightles with his uncle Vincent ploughing close behind him.

All the cows were 'down' again, reported Lambert, and blamed the rations, the waste carrots and the everlasting east wind, in that order. Fortunately the supplementary feeding-stuffs coupons had just arrived, so as a temporary measure extra cowmeal was issued all round, and Helen and Marina, whose milk yield had fallen to a most alarming extent, and in addition were 'off their feed', were dosed with 'Cordialine' from the veterinary cupboard.

The following morning I had to issue more 'Cordialine', this time for Snowdrop, who had broken loose overnight and begun an argument with Aldershot which had ended with Snowdrop spending several hours wedged in a bin minus one horn, and now needing first aid, a box to herself and general cosseting for a fortnight before her milk yield climbed back to normal. During the afternoon Old Tom appeared in the yard to announce that one of the heifers at Silford Hill was "a moanin' an' groanin'," and I rang up the Major, who told me to administer yet another Cordialine.

Between us Tom and I forced it down the heifer's throat. But I had to fetch Gidney, who was road-mending down

the Sandy Loke, before we managed to get her into a warm
well-littered box, where she settled down without apparent
discomfort.

"Was there anything fresh on the one o'clock?" Gidney
wanted to know, as we walked down the Loke. "Ha' we
started evacuating out a' Greece yet?"

"Not yet. But it looks as though we shall any day now."

"You know, miss, they ought never to ha' let the Germans into Salonika. Tha's where they done wrong," said
Gidney earnestly. "A place like that there. Whatever wor
they a' thinkin' on? I wor there t'ree year an' more las'
war," he went on unexpectedly. "Know it as well as I do
Silford. Aythens an' all. Though that worn't much a' place
so far as I could see. An' that Mount Olympus, where all the
fighting's going on now, there's a railway all round the
bottom a' that. Do I had a sovereign fer every time I ha' rid
on that, I'd be wuth something."

We finished loading the Doles carrots and I worked out
the crop and field averages. . . . Sixteen tons per acre sold
from Doles and an average of fifteen for the rest, the Prisoners Close reaching the highest figure with just under
seventeen tons an acre. There would have been more had it
not been for the constant frosts, which were occurring at
night even now; for 'waste', which when the Limekiln
Breck was worked amounted to less than three cwt. an acre,
had risen to nearly two tons before the end. Myhill, with the
air of a secret service agent who has unearthed a nest of
spies, handed me a metal ring from the leg of a crow he had
just shot, marked: "Vogelwarte, Rossitten, Germania.
D76436"; and it was all I could do to persuade him that the

bird had probably been ringed by some ornithological so-
ciety, and was not a sinister herald of coming invasion.
From Meddenham I fetched a store of assorted nuts and
bolts, the 'skimmer' which Cawston's had omitted to deliver with the new Midtrac plough, half a dozen wings and
three dozen points for the three-furrow, and another four
coulters for the corn-drill. On Three Bridges Roger drilled
the first of the sugar beet. Vincent reported that one of the
tracks on the 'crawler' was liable to 'go' at any minute, so
he and Mark spent a day round by the fitter's shop putting
on the new one, plus a new sprocket wheel, and reversing
and tightening the track on the other side.

"I should rather like to learn to drive a tractor," I ventured, as I found them trying out. "Is it very difficult?"

"I could learn you to drive it in twenty minutes. It's
learnin' how to use it what take the time," returned Mark
profoundly.

With Patfield I planted out some of last year's left-over
kale, swede, and sugar-beet seed in the garden, and on
flannel in saucers in the office, to make sure that it was safe
to mix off with a new seed before sowing. I ordered twenty
tons more superphosphate and filled the latest Land Fertility Commission forms for basic slag received. I posted off
the current Milk Marketing Board forms, and made en-
quiries about half price milk for Cattons' new baby. I ordered a year's supply of crude oil for the washer engine,
and two cwt. of assorted nails, nuts, bolts and staples for
the store, and rang up Weldon's for two new ball-bearings
and bushes for the crawler. I declared five more of the Silford Hill fat cattle for Meddenham market, earmarking

one for the Red Cross agricultural sale scheduled for the first week in May. Cawston's delivered the ribs for the Cambridge roll, sixteen small and six large discs for plough coulters, twelve dozen points, twenty-four wings for the double-furrow plough, three dozen hoe blades and a couple of 'eccentrics' for the manure drills. Meale and Wire top-dressed all the wheat fields with nitro-chalk, six stone to the acre. Kirk sowed sulphate of ammonia and basic slag on Prisoners Close, Fair Close and Richmond's Place. Behind him Roger drilled kale, mangolds and sugar beet, and put in the first six acres of peas at the Swains end of Ram's Breck. Vincent 'knifed' the glebe field destined for carrots, wearing the eye-shield from his Home Guard poison gas outfit to protect his eyes from the dust he raised. In Africa British forces continued to retreat. The evacuation of Greece began. Plymouth, Portsmouth and London were blitzed again. So was an East Anglian coast town, and several more unofficial evacuees came to me in quest of official billeting allowances and work on the land. The first registration of women under the new Employment Order took place. Not a person nor an hour must be wasted in the grim struggle which lay ahead of us, trumpeted the newspapers. Visiting Hill, who was harrowing over the top of the glebe wheat I found two men standing well out in the growing corn, engrossed with theodolite and measuring tapes, and enquired what they did. They were getting out the plans for widening the main road just here, after the war, they told me.

We were nearly into May, but there was little or no sunshine. And still the east wind blew, all and every day, nipping off the tentative shoots put forth by trees and hedgerows, stripping one layer from everyone's nerves, so that old quarrels broke out again and new ones began. Gidney could no longer be heard singing at his work. Bushell fell out with Meale. Isbell sulked. Peachey and Lambert had another fight and the latter had to be reprimanded again, first for coming to work late and then for coming too early, since the effect of the original rebuke was to make him get up an hour and a half before it was necessary one morning. He dressed as noisily as possible to annoy Marsham through the cottage wall, arrived at the cow-house an hour before time, and, having milked his own quota of cows, spent some thirty or forty minutes twitting Bailey and Long, who had turned up punctually at six o'clock as usual, with being 'all behind'. Hill and Kettering, already twice warned, arrived on duty straight from a prolonged session at the Hart a third time, and were dismissed from the Home Guard. Major Rivers, who is invariably the better for delicate handling, was as liable to explode as a stick of gelignite. Samuel Pepys scratched Mrs Heyhoe. Even Marsham wilted.

"An' that'll go on right through to June the twenty-fust," he said dispiritedly. "Do you see."

"Another eight weeks. But it can't!"

"That that will. Come in on March the twenty-fust, and that 'ont go till the longest day come. Do you see if that ain't right. Nor that 'ont surprise me do we wor ter hev a mucky summer, an' all, seein' as the ash ha' come out fust. You know what they say."

I didn't, so Marsham proceeded to enlighten me.

" 'Ash out afore oak, in for a soak. Oak out afore ash, in for a splash!' Or some ha' it 'there'll be plenty a' corn to thresh!' You pay yer money an' take yer choice."

The wind blew on. There was frost at night. The days were for the most part grey and sunless. There was too much rain and then not enough, so the land was increasingly awkward to deal with. But the work of getting the remaining fields ready for sowing went on. Field after field was ploughed up, harrowed and rolled, then knifed and harrowed again when the growing weeds could be discovered, looking like so many loose threads of white cotton, by kicking aside the surface soil. On fields like Prisoners Close and Booters there were patches which had to be gone over again and again, and still the soil refused to fine down, the best that could be achieved being a surface composed of small clods like marbles. Peachey harrowed the rest of the wheat.

"And when he's finished discing the glebe tomorrow, Vincent must start rolling the barleys," the Major announced as we walked back across Coney Hill. "He'd better begin here, using the crawler and the set of three tractor rib rolls."

I gaped at him. To drag sharp-toothed harrows over growing wheat had seemed an odd enough, not to say a risky, proceeding. But to send a heavy machine like the crawler (to say nothing of the rolls) bouncing over the top of delicate shoots barely three inches high and only a few weeks old, seemed completely crazy.

But corn is tough, apparently.

"Good gracious, no!" exclaimed the Major, when I suggested that most of the barley would be destroyed. "Do it good. And if you don't get these clods rolled in, you'll have the devil's own job to cut it at harvest."

Marsham announced that one of the two-year olds on the

Duck Hut meadow was shrinking, and "getting as poor as wood." So the Major ordered it up into the horseyard, where it was given a drink, had its feet trimmed for some undiscoverable reason, and turned loose again, when it soon recovered. I filled in another crop of forms, chiefly for supplementary stock rations and dealt with Oby's N.S. 100.A. which had been returned to me with a request to fill in full particulars of every employee on the farm, our individual ages, and an exact description of the work each did. (As only two lines were allotted for this saga, I rang up the War Agricultural Committee who said yes, they knew. The Ministry were like that. And would I return the information on a couple of sheets of foolscap.) I ordered two dozen boxes for the one-horse rolls and another four hundred pounds of Cyanogas. The order came in to load the Hunts-and-Gooch and Upper Common barley on rail, so I went down to Dennington station to see about sacks and trucks. The Meddenham Red Cross agricultural sale was a master one, reported Marsham, who had taken the afternoon off to attend. Farming ud be a different tale do stuff always fetch such prices.

"They whatchercall bid up fer that heifer a' ourn. Made about five-pun ten more'n she was wuth. I see a cowcumber sold fer t'ree shillun, then they sold it agen fer two-an-a tanner. An' some one hed giv' a swan. That come to a pund fust to, eight shillun the next time. An' then they put it up agen an' that made ten bob. Tha's the way ter do business, so long as you aren't a buyin'."

Gidney and Kettering went picking docks and foul grass off the glebe. Mark ploughed up Clap Close. Vincent rolled Barn Breck, Doves Close, Pound Close and Hunts-and-Gooch. Roger drilled another six acres of peas on Ram's Breck and clover, wild white, trefoil and rye grass over the barley on the Banwell Three Corner, Limekiln Breck and the Pightles for next year's hay crop. Then he began sowing sugar beet on the glebe with Ted Long leading. Ahead of them went Frank Buxton with a one-horse roll. Behind them Hill was harrowing in the seed with a pair of light six-baulk harrows.

Watching him I was bitten with the desire to take a hand at harrowing myself. It looked simple enough. A pair of harrows, each about four feet square, and a couple of horses. All one had to do was to keep the latter moving and the former following a reasonably straight course. I seized the plough-lines.

"Here. Look you out fer them wheel-marks, miss," said Hill anxiously, as I started off. "Do Roger 'ont be able to see where he's bin." And with the heel of one boot he hastily traced in again the line I had inadvertently har-rowed out. "You've got over too far to the right," he added, as I did it again.

I pulled hard on the left hand plough-line. I went on pull-ing, but Briton and Bunny continued placidly on their way unheeding, and out went another stretch of wheelmarks.

"That ain't the way," Hill told me, reaching for the lines. "CUBBERE," he adjured them, giving the left hand line the gentlest of tugs. The horses swung over to the left. "WHE . . . EE . . . ISH," he twitched the other line, and they veered right. He handed me back the plough-lines.

At the fence I was to stop to scrape off the foul grass with which the glebe still abounded, and which soon clogged the harrow teeth.

"Do you tu'n round fust, miss," implored Hill, as I began to pull up. "You mustn't never stop horses with their heads to the road. Do somethin' wor to come along they'd be off like flyin'."

Obediently I began to steer them round so that they faced into the field. Before they were half way, one harrow had managed to pile itself up on top of the other in some incomprehensible and unnecessary manner.

"Take a bit a' gettin' used to," volunteered Hill kindly as he helped me to disentangle them.

We scraped the teeth clean and straightened up the harrows. I jerked the plough-lines.

"Come on, Briton! . . . Bunny! . . ."

Neither moved.

"Briton! . . . Bunny! . . ." I repeated firmly, crescendo.

Both remained stock still.

Briton was one of the newly broken-in colts, and might conceivably not yet be used to his name. But Bunny had been Bunny for thirteen years and should certainly have responded to his.

"Bunny!" I shouted sternly.

Nothing happened.

"BOO-NAH!" yelled Hill just behind my left ear.

We started off again.

'Double' summer time came in, to the great disgust of everyone on the farm, while old Tom remarked that it was no wonder the wind kept on as it did; there'd never be no settled decent weather time there was all this foolery about putting clocks back'erds and for'ard. The Army came and fought another battle, but this one lasted only one day instead of three. And while the place was overrun with lorries and tea-drinkers as before, likely sites for weapon pits were merely indicated to an inspecting staff officer instead of being actually dug. Harold Peachey and Jeremy Peacock came on Agricultural leave, and started weeding wheat with Meale, Wire, Cecil Peachey and Bowes, on the glebe. Blue Peter and the two-year-old colts were moved on to the Fair meadow. Finding the kitchen boiler unlit as I descended the back stairs next morning, I sailed out into the yard in search of Croft, whose job it is to light it first thing, and found him outside the riding-stable with Patfield, Peel and Gidney, who should have been chaff-cutting in the haybarn.

"I must see about hiss first," exclaimed Croft unrepentantly.

I had often caught him crooning sibilantly to Amy when washing her down, as though he were grooming a horse. For a second I wondered if he could be referring to something of that sort.

"Rupert Hiss," Croft continued excitedly. "He's landed in Scotland. It was in the seven o'clock. Hitler's head-cook and bottle-washer," he elaborated, as I continued to stare. "Hiss or Hess or whatever it is. I reckon the war'll be over next week, miss."

All day the farm seethed. One school of thought supported Croft, and anticipated an armistice before the month was out. Another advanced the theory that the Deputy-Führer had arrived slightly in advance of the invasion, so

as to be ready to relieve the king at Buckingham Palace. A
few were in favour of the more than likely, but less enthral-
ling solution, that he had merely foreseen a purge and was
attempting to save his skin. But there was only one opinion
as to what should be done with the uninvited guest and
Peacock voiced it.

"I'm vengeful, I am," said he. "I reckon he should be
give over to the Poles, an' no questions asked. Tha's the
proper place fer him."

Cows and horses were turned out to grass. One of the
former overbalanced into the river and it took three hours
to get her out again. Captain had a bad attack of colic, from
which he failed to recover, so once again the Meddenham
horse-slaughterer had to be sent for.

"And Brisk. He'll ha' to be goin' soon. Jeffries, he ha'
give that leg a' his up. An' do we go on workin' him same
as he is, we'll ha' someone complainin'," Marsham reported
a day or so later. "How much did they give you for Captain?"

"Thirty shillings. Why?"

"Course Patfield, he wor sayin' somethin' about an advertisement what he see in the paper, offerin' to buy old hosses
by live weight. According to that Brisk shud fetch about
twelve to fifteen pund, where Sykes, he 'ont be good for
more'n twenty-five bob by the sound on it."

I looked through the 'Agricultural' column and found
the advertisement.

"Better give them a ring," the Major advised. "We shall
be round your way sometime towards the end of next week,
if that will do," said the voice at the other end of the wire.

So Brisk's fate was sealed. Meanwhile he was turned off on the Walnut Meadow.

Roger drilled ten acres of Crabb's Castle and six on Lease-pit Breck with kale. Myhill and Evan Gidney gassed the glebe fences again. Isbell began clearing out the upstream drains. Kirk sowed sulphate of ammonia on the other strip of Prisoners Close and potash nitrate on Hunts-and-Gooch. Myhill set the clock gun on Ram's Breck and prowled round with his gun to keep the pigeons off the peas. Easter Field and Doves Close were harrowed and rolled. Leslie chain-harrowed and rolled the Three Bridge Meadow. With the departure of Patfield for three days and Bailey and Old Tom for four (being stock men), the last of the holidays were worked off, except for Peachey, who had elected to 'stay with his hosses' as before, Kirk, who wished to take his at Whitsun (when his small daughter would have hers), Marsham and Roger, who were taking theirs in odd half days again, and Mark, who could not be spared yet. I spent a week-end in London. And Major Rivers, who never fishes at home, suddenly announced his intention of spending a week or ten days beside a friend's trout stream in Wales.

A day or two before he left we went over all the unsown fields to decide what must be done whilst he was away.

"Of course, everything depends on the weather," he said, as we took refuge in the shed on the Three Bridge Meadow from a hail storm, which had suddenly broken over us as we were looking at the sugar beet. "You ought to have been A-hoeing by now. But nothing will grow with this wind and these frosts. Still, you must get the horsehoes going at the first opportunity. And Roger should be able to get in the rest of the kale, the next lot of peas, and start drilling carrots before I get back. He can put the swedes in as well, if you can manage to get the land ready in time. But the latter's more important than the drilling at the moment. You've got a lot more to do to most of the carrot land. Barn Breck and Pound Close must be harrowed again, directly you have had enough rain . . . about twelve points should do for them . . . and then rolled down with a one-horse roll."

I scribbled hastily in my note-book. Outside on the meadow hailstones as large as peas and robins' eggs were falling.

"The Allotments, Hunts-and-Gooch and the Upper Common will have to be knifed again. So will that glebe field next Pightles. When you've finished discing the savoy ground, it'll have to be left on the Cambridge roll until it rains . . . it'll want a good twenty points before you'll be able to harrow it again. The same applies to that bit of Booters that's for swedes. But Horse Close mustn't be touched unless you get about twenty-five. And don't attempt to do anything to it then if it looks as though more rain is coming. It's an awkward old piece of land. And if you don't happen to catch it just right, you'll never be able to crop it at all."

The day after the Major left for Wales it began to rain. By the following morning forty-six points had fallen.

"What about letting Roger harrow Hoss-Close this afternoon?" suggested Marsham, soon after breakfast.

"The Guv'nor said it didn't want more than about twenty-five points. And we've had nearly double."

"I ain't bin up an' hed a look. But do that wor to get out fine, that ud soon ha' dried up too much agen."

I looked up at the sky. At the moment there seemed every prospect of a miserable day. It was dull, grey and cold, with a distinct tendency to mist and fog. But of course it was still only half past six by Greenwich Time. We decided to wait for an hour or two to see if the sun would come out.

At eleven o'clock, double summer-time, there was still no sign of sun. The mist still hung about. Leaving Amy on the main road, we paced solemnly out into the middle of the field.

"Too wet!" I said with sudden conviction, having arrived at this conclusion, not after prolonged thought or by any conscious process of reasoning, but following the promptings of a strange new instinct which I appear to be slowly developing where the farm is concerned; so while I am still guessing most of the time, increasingly often I guess right.

"I don't know that you ain't right at that, now I ha' hed a look," Marsham concurred thoughtfully. "That 'ont take any harm time this mist's hangin' about, any road. We'll ha' ter see what it look like tomorrow."

Overnight another five points of rain fell. The day which followed was as damp and misty as the one before.

"Mr Jordan, he wor harrowing that old wet land a' his down yesterday," vouchsafed Marsham, as we had another look at Horse Close, and I still hesitated.

Marsham had a lifetime experience of the land. But even Marsham was apt to make mistakes, and, according to the Major, getting on to the land too soon was one of them. Yet the Major himself went in for lightning changes of front. Ten to one that he'd have had the harrows out by now had he been at home. On the other hand this might be the tenth time.

"It can't take any harm with this mist if we leave it another day."

"As you like," said Marsham amicably. "Then Roger hed better drill them there peas along the roadside a' Booters and the four acres a' top a' Clap Close."

Next day it rained again. It rained on the three succeeding days. So after all we had done the right thing by Horse Close.

"A' course, Mr Jordan, he ain't a heavy-land farmer," Marsham let fall during the course of the week. "He don't plough deep enough for one thing. I hed a look at some a' his chaps as I come past that way yesterday forenoon. Four inches ain't a mite a' good on heavy land. I lay you a shillun the Guv'nor'd ha' kicked our backsides, an' sent us off home do he'd ketched us ploughin' like that."

The Duke of Aosta had surrendered at Amba Alagi. The Germans were invading Crete. A Messerschmitt chased a homing Blenheim over the Manor, nearly removing the chimney-pots in the process. Tracer-bullets flashed through the darkness like fireworks; but it was the Messerschmitt which crashed on its way to the coast. In spite of all my form-filling, Oby was called before a medical board, and warned to report for military duty in a month's time. Walter Bushell, Noller and Bailey registered. H.M.S. Hood received a direct hit in her magazine and went down almost instantaneously in the Atlantic.

"I ha' got a couple a' nephews on har," volunteered Patfield in the yard the day after. "My missus' brother's boys.
Acourse we hain't seen nothin' on 'em since Bob died in the
last war. His wife, she worn't wuth nothin', I'm sorry to
say. So we never kep' up with her, once she went off an'
took 'em with her. Still, it bring it home to you, don't it?"

"How old would Freddy be now?" asked Marsham, who
was standing by.

"He's worked his way up, you know. He's an engineer, so
Mrs Peel was telling my missus. She hear from Bob's wife
now an' agen. He'd be about thirty now."

"Tha's right, I recollect Roger was just beginnin' to run
about when Bob got married. Jack, he'd be a year younger.
Tha's all a job."

"Tha's right," agreed Patfield sombrely.

Peachey began A-hoeing the sugar beet on Three Bridges,
with Evan leading. Peacock mended the gate and railings
on the Fair Meadow broken by the colts. Mark overhauled
the roadwork Fordson. Bushell spring-cleaned the carrot
washer. Hill harrowed Doves Close and Kirk rolled it behind him. The house-martins arrived and thronged the
electric light and telephone wires. Roger sowed another
four acres of peas on Booters and started drilling carrots
on Barn Breck. The east wind still blew.

"Mr Nicholas done a good job when he planted all them
larches alongside the Chutch lane on Pound Close, an' run
that plantation out inter this here one," remarked Marsham as we crossed the field to speak to Roger. "I ha' bin
drillin' on here when half the field's been blowin' over into
Ashacre, an' all the carrot seed along with it, when the

wind's bin same as 'tis today. Tha's aburyin' the seed down yon end now as 'tis."

We walked on up the Sandy Loke to look at the seed kale on Mautby's Hall, now a sea of yellow bloom, with the pungent, none too agreeable smell. And then went down the hill and on to the far end of Crabb's Castle to where the new crop of fodder kale was coming up in neat straight rows. When we had inspected it only two days before, the heart-shaped seed-leaves had been fresh and green. Now a number of them were disfigured with tiny biscuit-coloured scars.

"Flies!" exclaimed Marsham. "Look at 'em, the mucky varmints."

Sure enough, there they were, the little yellow striped black beetles which had spoilt the six acres of kale on Leasepit Breck last year, hopping about like fleas.

"Tha's all along a' us not being able to get the land fined down properly. They can hide in amongst them there clods," went on Marsham. "There ain't a lot on 'em on this here yet. But I reckon we're in for some trouble with the flies this year afore we ha' done."

We went on to Leasepit Breck, where kale was coming up on the acres where barley had been last year. Here the 'flies' and the scarred leaves were more numerous. On the bottom of Lower Prisoners it was worst of all, though all the kale had been fly-free less than forty-eight hours ago. Kirk had to be stopped rolling Doves Close and sent off to Lower Prisoners instead, dragging paraffin-soaked sacks behind the roll, and re-soaking them at every round. Cecil Peachey was despatched to Leasepit Breck on a similar errand. Next day he had to move on to Crabb's Castle.

The Major came home from Wales, having failed to hook a single trout, but bursting with all the 'inside information' he had netted. We got the Bismarck. The shadow of coming evacuation hovered over Crete. President Roosevelt, in proclaiming a state of 'Unlimited National Emergency', gave details of the recent British merchant shipping losses. The newspapers were full of the seriousness of the food situation. The buyer of horses, who had so far failed to call at the Manor, suddenly arrived late one afternoon and offered thirteen pounds for Brisk . . . or we could have him weighed on the weigh-bridge at Norwich, and receive the advertised price of a pound a hundredweight.

I rang up the Major.

"We'll take the thirteen pounds," I said, as I put the receiver back.

"Right you are, lady. If you'll just lend me a pen. . . ."

"The advertisement said 'cash'," I remarked negligently. "After all, I know no more about you than you do about me."

"Anything you like, lady. If you can change this . . ." For the first time in my life I was confronted with a fifty-pound note. "Or you can give me a cheque for thirty-seven pounds."

Neither of these alternatives commending itself to my (unnecessarily, as it transpired) distrustful mind, he wrote out a cheque for thirteen pounds and promised to call for Brisk early the following week when the cheque was cleared.

"I couldn't have taken the old hoss today in any case," he said amiably. "I ha' got my lorry loaded up with four now. A beauty one is, too. Not more'n eight year old. But it's broke its leg, so it's no more use to the man what owned it."

"What are you going to do with them?" I asked curiously, as he folded up his cheque-book.

"You'd be surprised what gets into meat pies these days," he told me with a grin. "Not but what I'd just as soon eat horse as some a' the cows there are about. Horses usually cut up clean. I've opened up a lot lately, and only had one bad 'un among three hundred. You want to see some of the T.B. cows I get hold of."

It had been a perfect day, sunny and windless. After lunch I had actually put on a cotton frock for the first time this year. Next morning the wind was screaming over the fields again. It was icy-cold. Soon it was pouring with rain. Peachey and Hill had to stop A-hoeing sugar beet and went muck-carting instead. Roger left off drilling carrots on the Upper Common and 'jobbed'. Patfield sorted over the empty pulp and manure sacks. Catton, Noller, Gidney and Old Tom mended roads. Will Long helped Peacock in the carpenter's shop. Mark started his holiday. Every one else went weeding barley, muck filling or fencing. Marsham was oddly subdued all day. When I wanted to know if he felt ill, he owned up to another 'attack'.

"Look here. If you won't promise me here and now to see the doctor within the next few days, I shall ring him up and tell him to come and see you. And that's that."

"I don't know as that 'ont be just as well to let him run his rule over me," Marsham agreed, avoiding my eye. "I'll bike over a' Sunday morning."

"And you'll tell him exactly how you feel this time?"

"I might as well."

Next morning I passed on the news to Mrs Heyhoe.

"As if he don't know quite well as the doctor 'ont see anyone a' Sunday unless it's urgent," said she indignantly.

I went in search of Marsham and delivered an ultimatum which he received with a surprising meekness, promising to be ready to start at half past nine the following morning. Then I tried to ring up the doctor for an appointment. I tried again three times before I got Amy out next morning. But though the exchange connected us and I could hear the bell ringing at the other end, no one answered. It would be too infuriating if by some mischance the doctor was away when I actually landed Marsham on his doorstep.

"No. A barrage balloon broke loose and has fallen across our wires," the doctor told me, when I pounced on him for a word as Marsham, unfamiliar in a suit of black broad-cloth and a black trilby hat, disappeared into the surgery looking like a criminal walking into the condemned cell.

I went back to Amy and waited. The minutes passed. Other patients sidled down the drive and into the door. Then Marsham suddenly appeared, radiant, looking ten years younger with relief.

"That ain't what I thowt 'twor at all," he cried jubilantly. "Seems I ha' got a touch a' the same trouble what me feyther used to hev. Something wrong in one ear. Tha's what send me all over giddy now an' agen. I hev to go a bit slow-like, course I don't get no younger, like everybody else. An' I'll ha' another go from time to time. But that ain't nothin' to speak on. Me feyther he went on hevin' 'em for years, an' they din't do him a lot a' harm. I feel pounds better already."

So did I, when I had had a second word with the doctor, and secured corroborative evidence. Our journey home was in the nature of a triumphal procession.

"I don't know as I ha' got a lot agen doctors when it come to it," said Marsham thoughtfully, as I dropped him at his gate to change back into his working clothes.

Roger finished drilling the Upper Common. Hill began A-hoeing on the glebe, and Peachey started to sidehoe Three Bridges. Kirk sowed superphosphate on the top of Booters for the swedes and Wire harrowed in behind him. By pure chance I was moved to ring up the dairy outfitters and ordered a couple of new overalls apiece for the cowmen two days before clothes-rationing was announced. Two of the semi-wild ducks in the yard, whom I had noticed disappearing into the shrubbery beside the drive from time to time, suddenly arrived on the pond, each with a flock of squeaking black and yellow babies. Later in the afternoon, in search of decoration for the flat, I broke off a branch from the barberry bush behind the laurels at the end of the tennis lawn and came on what at first sight appeared to be a large oval ball of moss and lichen which blended so perfectly with the bush that it was invisible a yard away. Further investigations revealed a tiny hole half way up the side facing into the prickliest part of the bush. There was a scuffling and flutter of wings, and out flew a minute black-capped, black-backed bird with a tail quite twice its length. Perching on a rose bush a few feet off, it waited anxiously for me to go away, the while its tail twitched agitatedly up and down.

"Tha's a long-tail tit," Mrs Heyhoe told me next morning. "You don't often come across one a' their nests. I hain't found one since I wor a child, when I happened on it in a gorse bush on the common. I put me finger in to see if there wor any eggs in it, an' the bird wor on, and that didn't half lay hold an' give me a nip. Do you fancy a few stewed gooseberries over the week-end, miss?"

"Gooseberries? Has Croft brought any in?"

"Not him! Leave it to him an' you'd never hev nawthin' out a' the garden until that worn't wuth eating. He's like the rest on 'em. Can't bear to pull anything he's growed. So I ha' bin an' got a few on my own. Croft, he'll mob an' go on do he see 'em, I daresay. But you always look to hev gooseberries at Whitsun."

It was the last day of the coldest May for nearly forty years. The kitchen was warm and cosy as I sat at the edge of the table, helping Mrs Heyhoe to top and tail the gooseberries. Outside, the chill east wind still blew.

❧ XII ❦

YOU want to drill kale and swedes so's they come up when the elderberries are in bloom, by rights," Will Long informed me when I paid him over the money for twenty-two acres of chopping out at thirty-eight shillings an acre, on Friday. "They draw the fly off the fields, so my grandfather used to say. He always drilled his to come up then, an' he never lost a field a' either, all the years he wor at work."

Twice and three times a day during Whitsun week I had ranged over the kale on Crabb's Castle, Leasepit Breck and the Prisoners on the look-out for 'flies'. Often in the morning there were none to be seen; but by midday they would be on almost every plant, and I would have to order off the one-horse rolls at once and take a trailer load of tractor paraffin round with Amy.

Twelve points of rain fell Friday night. On Monday there were another twenty-one. On Tuesday fourteen fell. So for a day or two the flies suffered something of an eclipse. Then the wind actually dropped a little. The sun came out. So did the elderberry flowers. So did the flies. Soon Leasepit Breck had succumbed altogether. Gaps grew in row after row on Crabb's Castle and the Prisoners. They swarmed all over the Horse Close and Booters swedes hardly before the plants had had time to push their way above ground.

"Never see such a job," complained Marsham. "Do the fly once come in ahead on you, seems you can't stop the buggers. I mind one year as the Pightles wor mangels. Fly don't touch them as a rule. Do tha's a different sort what hang underneath the leaf so you can't get at 'em. But somethin' got that lot. Took 'em off, every one, as clean as a whistle. We never did find out what t'wor. So we hed ter break up the land agen, an' I drilled it with swedes. That wor flies hed that all right. Not a row left. Over she go agen, an' in with white turnips this time. An' would you believe it, if the flies din't tek them off an' all. So we hed to shuffle it up agen, an' put it in with mustard in the end. Afore we know where we are, we shall be on the same road agen t'year."

The seedling plants were rolled so repeatedly that they began to suffer more from the remedy than the disease. No more rain fell. Soon Marsham was beginning to pull a long face and talk about the drought. In the continuing sunshine the flies were as sprightly as crickets, and grew more voracious every day.

"I got talkin' to one a' Mr Bidewell's men down to the 'Ringers' las' night. He reckoned as they ha' bested their fly with basic slag," Marsham reported one morning.

"How much did they use? And how did they put it on?" asked Major Rivers promptly.

"He never let on nawthin' about that."

"Better run up to Bidewell's and find out," the Major said to me. "Turn left by Ashacre school instead of keeping on to the Hall, left again past the blacksmith, right at the top of the hill, and you'll find Bidewell's on your left about half a mile ahead."

I found the farm easily enough. But though a dog chained in the yard was raising echoes with his bark, I knocked on both front and back doors without result. The yard too, was empty. Turning the corner of a barn, I came upon a cart-shed in which three males of the species (father, son and grandson at a rough guess) were grouped in a trance around a tumbril.

"Mr Bidewell?" I murmured.

The youngest member of the party came to first.

"Dad, you're wanted."

With the portentousness of the world turning on its axis, Dad revolved round to face me.

"Ar?" he said inquiringly.

As I explained my errand, he unbent with amazing speed. Soon we were all being farmers together, companionably and profanely . . . it is surprising the vocabulary one acquires on a farm and uses in a matter of course, entirely without prejudice or malice aforethought . . . exchanging opinions on the weather and the flies. As one man to another, Dad asked my advice about the best width to drill carrots. Patiently he repeated the quantities of basic slag per acre and the time and number of men required to sow it by hand, whilst I wrote hastily in my notebook. Unfortunately he had only four acres of kale altogether. To carry out his method on the Manor farm would mean using over twenty tons of basic slag and necessitate about sixty men working at once, since the basic slag must be scattered by hand over the plants whilst the dew was still on them, unless it should happen to rain.

"Then that don't always do the trick," contributed the eldest generation, with a sly glance at Dad.

"They'll have to take their chance, that's all," the Major said when I got back to the Manor. As far as the kale is concerned, Roger can redrill Leasepit Breck and those bad spots on the Prisoners. And you'll have to top-dress all those that are left. A hundredweight of sulphate of ammonia to the acre will be enough, as long as you see the men scatter it only on the top of the rows, and not all over the place. It's the swedes I'm worried about."

"You aren't the only one," observed Marsham morosely. "Do there's any redrilling a' swedes to be done, that 'ont do to put that off much longer, do they'll never come to no size at all. I dunno what to think. Whether tha'd be better to hev 'em up now, or wait an' see if they'll get over it. That lot atop a' Booters, they ain't hed sich a doing-to as the others; they might just about get through. But Hoss Close, I dunno what to say about."

"You can write Horse Close off as soon as you like," returned the Major promptly. "They're done in. Mark must take the rotor up there first chance you get. But whatever you do to it, it'll never be ready in time for swedes again this year. You'll have to put in half the Easter Field with swedes instead of savoys; and that bit of Clap Close those peas were on."

"If only this yere drought 'ud break," lamented Marsham.

The wind died down to a mere breeze which rippled through the barleys, ruffling the swaying corn into wave after wave like water in some strange green sea. The sun shone with ever increasing vigour. The drought continued,

the land growing drier and harder every day. Roger drilled
the last of the carrots and the first savoys. Peachey horse-
hoed the sugar beet on Fair Close, Richmonds Place and
Banwell Close, and then started on the Hunts-and-Gooch
carrots, putting in two to three hours overtime every night.
Cecil horse-hoed all the kale again and then moved into the
Upper Common. The chopping-out gang started on the
glebe, where the still abundant weeds slowed up progress to
such an extent that the price had to be raised another two
shillings an acre. (When singling time came, an additional
seven shillings had to be added, or the men would have
been out-of-pocket by doing it on 'taken' work.) Oby received his last pay-envelope and his insurance cards and
left to join the Army, remarking thoughtfully as he said
goodbye that that 'ud be a change, anyway. The last of the
kale was top-dressed and hoed by hand. The new Fordson
for road work, ordered by the Major some six weeks ago,
turned up at last, and Mark fitted it up with the rubber
wheels from the one we were trading in. Deciding that it
would be as well to order a new Case, too, the Major rang
up Weldon's, and was told that delivery this side of Christmas could not be guaranteed, though they would do their
best. The longest day arrived, and Croft, grinning from ear
to ear, appeared in the kitchen in mid-morning to announce
that he was a grandfather again.

"Got a grandson this time," said he. "And I'd reckoned
on another granddaughter. I ha' got sucked in properly. I
said do that wor a boy, I'd buy him his fust suit."

"Look as though you'll ha' ter speculate a bit," Mrs Heyhoe told him, neighing with laughter. "Do you hev ter give

up some a' your coupons for it? Or has he got a ration book yet?"

For days the news had been full of a number of things. President Roosevelt had given all German consulates in the U.S. their marching orders. Mr Menzies had reorganised his cabinet. A year had gone by since France capitulated, and to mark the anniversary Marshal Pétain broadcast to the French anent a fuller collaboration with their conquerors. Germany and Turkey signed a non-aggression pact and spoke of 'mutual trust'. A U-boat sank the 'Robin Moor'. The Free French Forces occupied Damascus. Then the curtain went up with dramatic suddenness on yet another scene in that crazy comedy-tragedy entitled 'Change Partners' or 'A World at War'.

The day after Croft's grandson was born, Germany invaded Russia at four a.m. At nine p.m. Mr Churchill in a broadcast speech announced that the Allies would give whatever help they could to Russia. Newsreels depicting the U.S.S.R. at work and at play were shown in British cinemas to cheering crowds. The B.B.C., less certain of its ground, decided that the time had come to abandon its Sunday night musical medley before the nine o'clock news.

"Well, I dunno!" was Marsham's comment on the general situation. Nor did anyone else.

The drought went on.

On the hilly parts of Doles and Crabb's Castle, large patches of barley faded to a sickly parchment-white. From the top of Silford Hill similar patches, or 'scalds', could be seen in other farmers' barleys on the sloping ground beyond the river. On Ram's Breck the peas began to wilt and wither. The stunted pods ceased to swell. Inside them the growing peas lost their fresh sweet flavour and began to taste 'old' before they were within even hailing distance of maturity. The four acres of peas on Claps Close had been destroyed by pigeons a few weeks before. The new crop of swedes came up and a fresh army of flies at once descended upon them. But even the flies were beginning to lose heart.

Only the weeds thrived. Thistles poked their heads up above the glebe oats and barley. The horsehoes and scratchers were kept going all day, and an old horsehoe, long disused, dug out and repaired by Peacock, and started off again. For no sooner was a field finished than more weeds had sprung up, and it needed to be gone over yet again. Allotments, Hunts-and-Gooch, Upper Common and Barn Breck were a dark brilliant green, not with carrots, but with 'fat hen'. I wrote to the manufacturers for another two dozen crawler hoes, and the men as well as the women had to creep up and down the rows, hoeing the baby carrots clear.

Haying, though nearly three weeks later than last year, was upon us with the weeding and chopping-out still in arrears.

"An' tha's a werry moderate crop at that," grumbled Marsham the day before we started cutting. "Pity there worn't more clover in it t'year. That always stand up better to a drought."

"Harvest will be late too. There'll be no corn cutting on August the first this year," remarked the Major that afternoon, as we walked past the Upper Prisoners wheat, after inspecting the kale and mangolds on the lower field. "Have you noticed how the wheat has altered during the last two or three days? When wheat comes into ear . . . not just one or two, here and there, as it was on Monday, but the whole field, which comes on all at once within a couple of days . . . you know that harvest will be another six weeks."

Next morning hay cutting began, with Mark on the Case, Vincent on the cutter and Roger sharpening knives. Half way round for the second time, and they were in trouble with the cutter. Mark tinkered with it for a moment or two, and they started off again. Another two rounds and once more they stopped.

"I don't reckon as that cutter bar is set right. Tha's too low," observed Marsham, as once more Mark and Vincent unscrewed and screwed up nuts and bolts, and readjusted knives, keeps and fingers.

"Do that went all right like this last year," returned Mark as he straightened himself up. "Let's see how she'll do now, Vincent." He climbed back to the seat of the Case.

"They'll be stopped agen afore long," said Marsham, as they bumped off. "I know that bar ain't right. I'm off back into the yard to see if I can't scheme somethin'. I believe I kin lay my hand on a bit a' chain an' a pole off the old cutter as ud do the trick."

Before he was back again, the cutter had jibbed again. It was still stopped when I left them to go into Meddenham to fetch the crawler hoes, which had been consigned to Meddenham station by mistake. When I revisited Sidegate Breck in mid-afternoon, the cutter had been working continuously for the past two hours and was still cutting away.

"Old timers know somethin' arter all, fare like," grinned Marsham as we watched them take a corner off.

Hay cutting had begun on the Thursday. On Friday

morning the last blade was down in the blazing sun, and Mark and Vincent went off to help with the Ashacre Hall cutting. Saturday was dull and misty. There was even a threat of rain, but this was never implemented. Roger drove the toppler, a gang of men put up one row of haycocks, Evan followed them with a horse-rake.

First thing on Monday morning the hay was 'toppled' again. By ten o'clock the dew was gone, and every available man was up on Sidegate Breck cocking hay. At seven o'clock on the Wednesday morning I watched the two elevators being set in place. Will Long, Vincent and Roger re-adjusted belts and fiddled with the engine, while the pitching and stacking companies waited expectantly, forks in hands. Out in the field Mark took the Mugleston for a preliminary run round. Marsham, who had driven up with Jack and a cart-load of petrol, water, and paraffin, hitched him to the gate and ranged himself beside me.

There was a pregnant pause. Then Vincent and Will gave the engine wheels a final turn, there was a whir and a clatter, the elevators revolved, and Hill and Catton flung on the first forkfuls of hay.

"That ain't sich a bad crop, arter all," vouchsafed Marsham at midday. "Tha's going to be a master stack afore they top it up. But tha's how it shud be. Hay always do best on a big stack, arter tha's 'made' on the cock. Them mucky little stacks don't come to nothin' like sich good hay."

The stack rose foot by foot. The elevators were cranked higher and higher. Frank and Evan horse-raked behind the Mugleston. Just before eight o'clock the last wisp of hay was scraped off the field and carried up on to the top of the stack. From start to finish haysel had occupied barely a week.

"That there sweep certainly do push things along," said Mark, as he disconnected the Mugleston near the gate.

"As long as you get the right weather for it!" amended the Major, who had called round to see the finish on his way to a Home Guard meeting. "We've been lucky with it these last two seasons. But I've seen an eighteen-inch growth of second-crop in the field with the cocks still standing, when there's been a wet haysel. The Mugleston would be useless in a case like that."

At the moment it seemed unlikely that it would ever rain again. Sidegate Breck had grown so hard that the Mugleston had broken three teeth and shattered a connecting bolt during the day. The grass on the pastures turned brown, dead and dry. As feed it was useless, and the milk yield fell accordingly. The carrots on the glebe came up, but with gap after gap in every row, and whole rows missing on the hillier parts. The scalded patches in the barleys spread. Roses in the garden bloomed and died in a single day. The savoys came up on Doves Close and the Easter field lost interest, as though they had exhausted their entire store of vitality in making their way up through the unyielding ground. The strawberry crop was a failure. "So'll the raspberries be, do we don't soon get some rain," said Mrs Heyhoe dolefully. "I dunno, I'm sure." The Booters peas, already stunted and deformed, ceased growing altogether. On Ram's Breck the first-sown peas were beginning to die off at the roots, the lower leaves turned to parchment, the fresh dark green on the pods to a sickly yellow.

"Whether they're fit to pick or not, they'll have to go within the next few days," announced the Major. So I rang up Barretts and arranged for lorries as soon as they could send them . . . a matter of four days, since they were busy with Air Ministry work, they told me, and would probably have to hire lorries from other transport companies.

It grew hotter and hotter. Occasionally clouds obscured the sun, but they only made the atmosphere more sultry and oppressive. Tormented nearly to insanity by the heat and the flies, the cattle on Three Bridges broke out one evening, and Hill, Catton, Isbell and their families spent three strenuous hours getting them in again. Somewhat late in the day, but none the less welcome, the permit for the extra haysel rations arrived, and Arthurton's packed up infinitesimal parcels of tea, sugar and margarine, which Marsham and I distributed with the help of Jack and the cart. R. A. Leeds called round and gave me an orange, a somewhat depressing forecast for the pea trade, and the inside story of the latest Government marketing muddle. The men finished weeding the Upper Common carrots and went on to finish cleaning Hunts-and-Gooch, while the women left off to pick peas at the top of Ram's Breck.

"They 'ont be able to pull 'em for one-and-three a bag this year," said Patfield, as we walked over a cross-section of the field in an effort to size up the crop.

"I know. The Guv'nor said we'd better try one-and-six, and see how they get on."

"I doubt if they'll be able to do it for that, either," returned Patfield. "Bin up here an hour already, an' not one

on 'em with nothin' nigh a skepful yet. Look at this here."
He dragged up a plant, and stripping off the few meagre
pods, held them out to me. "Half the number a' pods there
ought to be, an' a quarter the size. They're beginning to
grumble already; an' you can't blame 'em. They aren't earning hour's pay."

At half past eleven, having first rung up Major Rivers, I
went back to the field, raised the price for picking to two
shillings per bag, and drastically reduced the consignment
list.

"An' when you get back into the yard, if you'd see Lambert bring a churn a' water up for us to drink," suggested
Patfield, as I left. "He's coming up with a hoss an' tumbril
to get a load a' pulled pea-plants for the cows this morning,
isn't he? Tha's somethin' dry."

By one o'clock it was obvious that even at two shillings a
bag the women were working at a loss. The price was
raised again, to half-a-crown this time, but several went
home when they had finished their first bag . . . which had
taken all the morning and part of the afternoon to collect.
Last season they would have picked five or six bags in the
time.

I reduced the consignment list again and rang up Barretts to warn them that only one lorry would be needed, and
a small one at that. The day's total eventually reached
eighty-nine bags. Next day's showed an increase of five
bags. The following day's dropped to seventy-eight.

The transport company whose aid had been invoked by
Barretts was making a fixed charge per journey for their
lorries irrespective of the number of bags loaded. So far the

transport costs were working out at nearly two shillings per
bag. A few of the first consignment made ten and eleven
shillings a bag, with one-and-a-penny commission to come
off. The rest made prices ranging from four shillings to
seven-and-sixpence a bag. Of those sent on the fourth day
of picking, some made three shillings a bag gross, some
half-a-crown, some were never sold. R. A. Leeds and A. L.
Gardiner, who had been telephoning and wiring us to load,
wired frantically, "Stop loading peas." Jimmy Bailey and
Frank heaped up the remaining plants on the top twelve
acres for Lambert to cart in for the cows.

"A master pity the pigeons didn't take the lot, time they
wor about it," observed Marsham sourly, as we ordered
Vincent off to Ram's Breck with the digger plough.

Overnight the weather turned dull, cloudy and cool. In
the early hours of St. Swithin's, it began to rain. Before the
day was out nineteen points had fallen. The parched land
absorbed it with the avidity of a scandalmonger lapping up
gossip. The savoys perked up. Tiny seedling carrots began
to appear in the gaps on the glebe. The fly-ravaged kale
and swede plants began to grow. The Booters peas took on
a new lease of life. We decided to give those on the bottom
end of Ram's Breck another chance.

Within the next three days another nineteen points of
rain fell, and a ferocious thunderstorm swept over the district, deluging the villages to the east of Wendeswell, but
missing Silford altogether.

"Do that hed ha' copped us, that 'ud ha' laid the corn so
we shouldn't never ha' bin able to cut it wi' a binder," Marsham informed me when, in my ignorance, I lamented our

loss of the rain. "An' don't you fret," he added, with a glance up at the clouded sky. "We shall ha' too much wet afore we finish, seein' as that started on St. Swithin's same as that hev t'year."

The sugar beet on the glebe and the Booters swedes were chopped out. Peachey side-hoed the Upper Common again. Cecil A-hoed the Lower Prisoners kale and mangolds. Roger rolled Ram's Breck behind Vincent and harrowed the still uncropped Horse Close. The chopping-out gang started singling the Three Bridges sugar beet, and the women left off carrot weeding and began to pick peas again.

The four acres of Ram's Breck yielded a total of sixty-six bags. Next day they began to pull the 'Onwards' on Booters, culling a total of eighty-four bags. The peas were well worth eating, fresh and sweet. But the pods were about a third of their proper size, and well under half full.

"Who's agoin' to spend all day coshing 'em?" demanded Patfield.

Nobody, it appeared, except at a price well below the cost of picking and cartage, to say nothing of something towards such trifles as initial labour and cultivation costs, and the staggering price paid for the seed.

"Better by half put the women back on to weeding carrots," said Marsham as we waited for the Major to come round next morning. "They'd be doin' a sight more good. An' arn more money too. Do they keep on wi' the peas, same they're adoin', much longer, there'll be half on 'em off to that aerodrome they're putting up on yon side a' Wendeswell. There's girls a' seventeen getting three or

four pund a week, so they tell me. An' that there Sam
ewacuee, he bring home four-pun-ten, so his mother wor a
tellin' my missus up on the pea field yesterday."

"Four-un-seventeen it wor last week, so he tell me when
he wor along a' me on Home Guard last night," chimed
Will Long, who had come in for the key to the black huts
to get out some nails. "An' I felt like tellin' him as that
wunt do no harm do he wor to buy some soap wi' some
on it."

"Tha's a fact," agreed Marsham.

The morning's post, which arrived on the doorstep simultaneously with the Major, settled the pea question out of
hand.

"Put the women back on to the carrots right away," the
Major ordered. "And Vincent can plough up Booters directly he's finished Ram's Breck. Anyone on the farm who
wants any peas can pick what they like before Vincent gets
there."

Soon after midday half Silford and Ashacre were picking peas on Booters. As the afternoon wore on they were
reinforced by streams of passers-by. A string of cars was
parked along the road fence with a dozen or two bicycles.
Two large policemen cycling by, alighted to see what was
going on and then joined in the fun. Rabbit keepers abandoned their hedgerow search for hog-weed, and carted
away sacks full of pea plants. The following afternoon Vincent moved in with the crawler and the digger-plough, and
began burying up the rest.

The women re-weeded the Upper Common carrots.
Peachey side-hoed those on Hunts-and-Gooch again. The

chopping-out gang finished the Easter Field swedes and transplanted savoys on Doves Close. Kirk sowed superphosphate on Ram's Breck, and behind him Evan rolled, Roger drilled white turnips 'soshing' (at an angle instead of up and down the field) and Frank harrowed in the seed. The seed-merchants, for whom we were growing seed-kale, wrote to know if we could undertake some acres of swede-seed for 1942 as well, which need not be drilled for another three or four weeks, so a possible crop was found for Horse Close. Mark began to overhaul the power binders in readiness for harvest.

"I hear you've just bought a car," I remarked, as I arrived in the barn to ask what spare parts were wanted.

"Tha's only an Austin Seven," began Mark defensively, as though he expected me to take it away from him if it was anything bigger. "My wife, she's bin on at me about one some time. Peel ha' hed one a good while, you know."

"Has he? Well, why not?" I asked curiously, unable to account for Mark's slightly hostile tone. At my question, his prickles subsided.

"You never know how people will take things," he confessed, his cheerful self again. "Some on 'em on the farm ha' bin talkin' an' turnin' the grin on me. What I say is, there's nothin' ter stop any a' them gettin' one an' all, so they like to give up some a' their baccy an' beer. I don't mind what they do wi' their money. But fares as though they can't rest do I want ter spend mine different."

"Personally, I should think you'll get a lot of fun out of it. I hope you will, anyway."

"I don't know about that," returned Mark primly. "But

that'll come in rarely handy ter go an' see me wife's mother at Gilderstone of a Sunday. An' ter go out fer a drive now an' then: there an' back, ter see how far it is. You'd better get some more links fer that chain, miss. An' we shall want some new fingers."

I wrote to the Food Office for extra harvest rations, the prospect of obtaining which had actually been announced far enough in advance for once to make it possible to obtain the rations whilst the work was being done. I was officially notified that our application for the retention of Oby's services had been considered, with the result that he would be left with us until October the first, which missive I endorsed, "Called up five weeks ago," and dropped back into the letter-box. The same post contained a letter addressed to "Mrs Nicholas Irstead, The Manor, Silford," which I handed over to the Major unopened.

"This is meant for you," said he, handing it back to me, with a broad grin, having first hastily skimmed it through. "Kate's got one this morning as well. You'd better call round and see her during the day."

Short and to the point, it stated that there would be a further large evacuation from a certain East Coast town within the next seven days, when "Mrs Nicholas Irstead" would be expected to accommodate a family at the Manor, at twenty-four hours' notice. During the morning half my women produced similar notices and besought my aid.

"Do they make us take in evacuees, we'll ha' ter give up working on the land," was the burden of their song. Two literally had no spare room, four were expecting relatives for the Bank Holiday week-end. All of them knew of houses in the village with as much or more room than they themselves possessed, where the housewives were at home all day into the bargain. Yet these had received no notices. Wasn't there anything they or I could do?

Nothing, except make the best of it and I'd do what I could in the way of re-billeting at the first opportunity, I had to tell them after ringing up the R.D.C., the local inspector and the Sedgeham W.V.S.

"You never know when we may be evacuees ourselves," I reminded them . . . and myself too; for already the days seemed all too short for the work in hand. And with harvest just ahead, I felt that the authorities could well have spared us further problems until the corn was gathered in. But there it was. Mrs Rivers spent an hour or so going through the house, deciding which room could be used, Mrs Heyhoe temporarily abandoned carrot-weeding, and between us another flat was made ready downstairs, extra beds, bedding and crockery were delivered by the W.V.S. (since the Manor's supply of both was hopelessly inadequate), Mrs Rivers unearthed an oil cooker, and Peacock and Will Long fitted black-out and Yale locks to the office door.

The weather, already cool, turned colder. We had nearly three-quarters of an inch of rain during one evening and night, and another thirty-one points twenty-four hours later. The Major, who was about to spend four or five days on Home-Guard manoeuvres, made a tour round the corn fields with Marsham and me, to decide if there was any likelihood of any corn being fit to cut in his absence.

"Them there oats on the glebe are pretty nigh fit to cut now," said Marsham, as we drove up Church Lane.

"Yes. Olland oats always ripen first," answered the Major. "If the weather holds fine, you might cut them on Monday, if Mark and Vincent will work Bank Holiday. And they can stop and cut the wheat whilst they're there. It's no good moving all the tackle over to Banwell Close and then bringing it all the way back again just for that bit of glebe wheat."

"No, that it isn't," agreed Marsham. "I still wonder if that wun't ha' bin best ter ha' ploughed that in fer a start. Tha's as full a' rubbidge now as ever that can stick. There 'ont be more'n about t'ree load when we come to cart it."

We waded through the offending wheat, scarlet with poppies and green with thistles, in spite of all its weedings, and paused on the edge of the barley field beyond.

"That ain't so bad as I thowt that was goin' ter be," observed Marsham, pulling up a handful. "That ain't far off ripe, neither."

"You know what they say. When you think your barley's fit to cut, take a fortnight's holiday, and begin cutting it when you get back," the Major told him. "You can get straight on to the Banwell Close oats directly you've finished the glebe. But there'll be nothing else fit to cut for another ten days at least."

The weather continued dull and cold. There was an autumnal tang about the mornings. A 'sea mist' hung about most of the week-end, varied by occasional small showers. On Monday morning the sun came out, and stayed long enough for Mark and Vincent to cut down the glebe oats. But before they had been more than once round the wheat, a sharp shower fell, and they had to stop and sheet up the binder.

For perhaps an hour they shocked up, while the sun

played at hide-and-seek with the low-lying clouds. Then a fine thin drizzle set in. Walking round the farm in the late evening, I was suddenly engulfed in a cloudburst and soaked to the skin, in spite of my mackintosh. On the way back I took off my shoes and splashed ankle-deep through the water in the hollow in the Ashacre Road.

Next day, though the temperature was reminiscent more of December than August, the weather was behaving like April. Brilliant sunshine alternated with lightning showers. Soon after tea there was a first-class thunderstorm, in the midst of which a car with a large W.V.S. across its windscreen, drove up to the front door.

"Is this Silford Manor?" demanded the driver, consulting a paper. "Two women and three children. But one of the women hasn't come."

The woman who had and three small children scrambled out on to the drive. "I don't want ter stay here," the three latter wailed as one child. "Do you don't hold your row, I'll give it to yer," their harassed mother proclaimed.

"Well, here you are. You'll be all right now. Miss Harland will look after you," the W.V.S. said brightly, ignoring all these noises, and promptly let in her clutch.

By the time we reached Flat II, the wails were almost deafening and Mrs Gaff was administering cuffs impartially, varying her threats on her own account with, "The lady'll smack yer bottoms for yer in a minute."

"Oh, no she won't," the 'lady' said hastily, and fled into the kitchen to make tea and scramble eggs.

As I reached Flat II with a tray, the front door bell rang, and I found Gidney on the doorstep.

"Please, Miss, there's three evacuees standing about in the

road up by mine, cause the people they're billeted on won't have 'em. At least, they wor standin' there, but I ha' took 'em in an' my missus is giving 'em a cup a' tea. But we can't keep 'em as we ha' got four shot on to us already."

"Right you are. I'll fetch them in a minute," I promised. Putting on a mackintosh I went out to get Amy, pouncing on Croft on the way and extracting a promise that he and Mrs Croft would house the unfortunates for a night or so whilst I looked round for another billet. When I got back into the house again, Mrs Rivers was on the telephone.

"There's a lorry on its way to you with your evacuees' pram and luggage. And I hear there's been a mess-up at Silford Hill. But Mrs Bellboddy in Port Row will take those evacuees in right away, if you like."

Since Croft was taking his holiday in two days' time, and going to Birmingham with Mrs Croft to see their grandson, it seemed simpler to deal with the re-billeting at once, so I got Amy out again. Back at the Manor once more, I knocked on the door of Flat II, and announced the safe arrival of the pram.

Kathleen (aged five) and Daphne (aged two) were now merely whimpering. Only Herbert (aged eight) continued to raise his voice unabated.

"He's just had an operation and had one kidney took away with a tumour on it," his mother shouted proudly above the din. "We han't half hed a job with Herbert. I 'ont never forget the day they took him into hospittle for the X-ray, and we hed ter wait ter know do that wor a male or female tumour."

"What did it turn out to be? Malignant?" I suggested.

"Tha's right. A malingering tumour. Tha's what Her-

bert hed. Do you don't hold yer noise, Herbert, the lady 'ont
half smack your face for you."

Altogether between twenty and thirty evacuees had been
deposited in Silford that evening. By noon next day quite
half that number had been down to the Manor to complain
about the lack of shops, cinemas, baths and indoor sanitation. And a new version of the Silford Hill episode was in
circulation, from which it appeared that the householder
had been welcoming enough, but the evacuees had 'turned
up their noses at the place.'

"Don't no one tell 'em what the country's like afore they
started?" demanded Mrs Heyhoe. "Mrs Davey, she tell
hers, we hain't baths, though we'd like 'em, she say. But we
keep ourselves clean just the same, an' make the best on it.
Come out a' council houses, so they say. I'd give 'em council
houses do that wor me."

Before another twenty-four hours had elapsed, several
evacuees had walked either to Banwell station or back to
Sedgeham and were on their way home again. Mrs Gaff
appeared to have quite settled down in Flat II, however.

"Some on 'em wor askin' me how I wor a gettin' on, this
morning," she told me on Saturday evening, as the family
accompanied me to watch wheat cutting on the Upper
Prisoners. "An' I told 'em as I wor doin' very nicely. I hed
sich a turn when I fust come. I said to meself, this is much
too posh. But tha's turned out quite homely. An' my husband, he said he didn't know the lady's name, but he
wished to be remembered to you. An' would that be all
right for him to come when he gets his leave? KathLEEN
come you off that corn, do I'll give you a clip a' the lug."

Mark and Vincent cut the beans, black-ripe, on Gypsies.

The clocks went back an hour. A deputation of his neighbours implored me to speak to Lambert again about his black-out, and I threatened him severally with eviction, the Major and the police. Kitty and Beauty started work again, and their foals, bereft of their mothers for the first time in their young lives, protested by kicking at the door of the box in which they were confined, kicking over the buckets of water proffered by Marsham, and snorting and whinnying unceasingly for several days and nights before they gradually began to grow accustomed to their loss.

Milk rationing was announced, and I rang up successively the Food Office, the War Agricultural Executive Committee, the Farmers' Union and the Milk Marketing Board, in search of information. Nobody could tell me anything definite. Hadn't I been listening to the wireless? No, I hadn't. Well, there *was* something about no registering for employees who received milk in lieu of wages. Did mine? No, but they soon could. Only in yesterday's paper there was an account of a body of farmers complaining because they weren't allowed to do that very thing. However, I could try. . . . So, to the complication of an already complicated paybook, I added milk allowances, and the men promised to come and bail me out should my efforts on their behalf land me in gaol.

Bertie Buxton and Herbert Bushell arrived on agricultural leave. The harvest ration arrived, eighty-three separate parcels of tea, of margarine, sugar and cheese, and everyone marched past the kitchen window while Mrs Heyhoe and I dished them out in sets of four. Large V's mysteriously appeared on all the wagons and buildings.

Officially we were up to harvest, but every day it rained, and such corn-cutting as was accomplished was done piecemeal, with one eye forever on the lowering sky. To cart and stack it was out of the question.

On the glebe and Banwell Close, the bright yellow oat straw turned dull and blackened, the grain began to sprout in the ears. On the Upper Prisoners some of the wheat in the sodden sheaves sprouted shoots nearly four inches long. The barley shocks on Coney Hill blackened like the oats. On Doles, as yet uncut, the 'scalded' patches turned a bright green with second growth. Last year the seed kale had been cut and threshed well before the end of July. This year it was to be cut nearly three weeks later, not until the seeds, fallen from the bursting pods, had sprouted and covered the field like cress. Again and again everyone stood by for carting and stacking, and then the rain began once more, and instead they went hedging, repairing, chopping out the turnips in Ram's Breck and Booters, or gave up altogether and went home.

"A master good harvest bein' buggered up by the weather. Tha's how 'tis," said Marsham disconsolately. "What we want now is a right good wind to dry all the corn on the shocks. An' I tell you somethin' else. Do we get it, that'll take half the ears off the standin' barley afore we can cut it, 'cause tha's rotten ripe now. Do I could ketch holt a' old St Swithin, I'd like ter wring his neck."

Though several more evacuees had drifted home again, the Gaff family appeared more or less permanently settled in Flat II. Mrs Gaff joined the carrot weeders on the glebe. Samuel Pepys adopted Daphne and Kathleen, and occasionally responded to the name of Muff-Muff. Herbert shyly confided one morning that he liked Silford, though he liked England(!!) best. Corporal Gaff arrived unexpectedly on twenty-four hours' special leave, during which he found time to give me his version of the family history at length, plus a new and enlarged edition of Herbert's operation, and promised me a box of red herrings cured with his own hands, the moment the war was over and he back in his own job.

The night following his departure, hearing noises below my window, I looked out. And below in the uncertain light and the inevitable rain, I made out a sailor standing in the drive.

"Please, could I see my wife?" he wanted to know.

"But my evacuee's husband is a soldier," I protested, somewhat stunned.

"This is Silford Manor, isn't it? They told me that I'd find my wife here," he persisted. "Willis, the name is."

It was one which figured on none of my lists. I rang up Mrs Rivers.

"Willis is the name of mine. But she is in bed and asleep. What do you want her for?"

"I've got her husband here. He's just off a minesweeper for forty-eight hours, and he's wasted nearly half of it hitch-hiking round Norfolk looking for her. I'd better bring him round, hadn't I?"

"Yes. And I'll rouse up Mrs Willis."

A. B. Willis had lost two ships and an elder brother. But like Alf, "you ha' got ter laugh," he said, as he described his last escape. "You know, having a wife an' family, tha's apt to make a child a' a man. But do he get back to livin' among men, that soon make a man a' him again," he added thoughtfully, as we drew up at the Hall.

The next few days continued showery, though with infrequent bright interludes. Then came a whole fine day. The Upper Prisoners was re-shocked in the morning, and carting started in the afternoon. Mark and Vincent finished cutting the North Field wheat and moved up to Mautby's Hall, while Roger and Leslie used the second power binder in Swains. At Wendeswell Hall a combine-harvester was at work. And after an unavailing search for Marsham to accompany us, the Major and I set off to Wendeswell to watch it in action.

Pulled by a bright scarlet tractor, the combine (silver picked out with scarlet and yellow) was sailing round a barley field, swallowing up the corn whole as it went and spitting back the straw on to the field, whilst the grain fell into waiting sacks. Only two men were needed to work it. One drove the tractor and, sitting at an acute angle on his seat, also manipulated the steering-wheel of the combine. The other stood up on the sacking platform at the side, taking off the full sacks, replacing them by empty ones, and sending the former sliding down a metal runway into the field, to be collected by a following tractor and trailer. Cutting, shocking up, carting, stacking and threshing were all achieved in the one operation.

As I got down from the platform, having ridden around, Marsham drove in through the gateway with Jack.

"I thowt as I'd draw round and hev a look," he said sheepishly when he caught sight of us. "Tha's a rum 'un, isn't it?"

"It is," I agreed enthusiastically.

"Though I can't say as I reckon much to 'em meself," Marsham went on, showing no enthusiasm at all, as we walked down the field behind the machine. "Look at the way all the muck is blowin' out over the field. Do there's some dock seed in it . . . an' I lay you a shillun there's a plenty . . . they'll ha' some trouble in that next season. Tha's whatchercall splittin' up the straw an' all. That 'ont be no more'n catmuck by the time they clear that up."

"I wouldn't have one as a gift myself," contributed the Major. "To do the job properly you want a straw baler following on behind. Then there's the question of drying your corn. Either you have to install your own drying plant or sell it to someone who'll take it damp, which means you're pretty well in their hands. And then there's the time and weather factor. If you were sure of plenty of fine dry weather, I daresay it mightn't be too bad. But she won't do more than about ten or twelve acres a day whereas you can cut and cart thirty in the ordinary way. I daresay they'll overcome most of the snags in time. Meanwhile they can keep their combines for me. I've done my share of pioneering. . . ."

Next day opened fine and sunny. By breakfast time wheat-carting on Upper Prisoners was in full swing. Whilst everyone was home for dinner, a sudden thunderstorm burst over Silford and Ashacre. The Major got the worst of it, but we had fifteen points of rain within half an hour. Then the sun shone out again. But in less than two hours another storm burst, and nineteen points cascaded down. Wheat-carting had to be abandoned. Cutting was out of the question. Isbell went cleaning drains. Mark and Vincent tidied out the fitter's shop. Peacock and Will Long repaired a wagon. Everyone else weeded kale or picked up fallen sheaves.

The following day it was comparatively fine. The Upper Prisoners dried off sufficiently for carting to start again. Roger and Leslie finished cutting the Swains wheat and did a couple of rounds on the Coney Hill barley. Vincent and Mark got the last of the Mautby's Hall wheat down just before eight o'clock.

"We got a rabbut or two up there an' all," reported Marsham, as we met in the yard just as I was about to walk up to Silford Hill to see how they were getting on. "I ha' just bin sharin' on 'em out."

"Were there enough to go round?"

"Only for them on the field. There don't fare to be any rabbuts about, hardly, t'year. But there wor just enough for one each fer them as wor there to-night, acourse I happen ter see Gidney knock a couple over, an' give 'em ter that ewacuee woman what live along a' them, ter tek home. So when it come to a share out, 'Joe,' I say, 'You ha' got yours, hain't yer?' Well, acourse, he hed ter say yes. An' that'll be one fer Hill an' all. He'll call in fer it on his way home, I tell him. Some a' the others might ha' gone home without it. But I knowed as Hill 'ud hev it orf him," grinned Marsham.

Nineteen points of rain fell during the following day, and everyone 'jobbed' or chopped out turnips on Ram's Breck. The day after, Sunday and the last of St. Swithin's allotted forty days, was a perfect summer day with a brilliant sun riding a flawless blue sky. But next morning it was raining again.

"I thought how it would be when I see the stars runnin' las' night time me an' Heyhoe come walkin' back from his mother's," said Mrs Heyhoe darkly, as she hung up her mackintosh. "You may always know there's more rain acomin' when you see the stars runnin' across the sky."

Before ten o'clock Croft appeared to beg for aspirin, thermagene cottonwool, and 'The Three Musketeers', after which he retired to bed with lumbago. All the women stayed away. Three parts of the men lost half a day. The other half they chopped out turnips, carted muck or straw, or went hedging. Noller, engaged on the latter up the Sandy Loke, inadvertently caught his hook in an unsuspected wasps' nest.

"He ain't half hoppin' about, neither," said Marsham in his best Cheshire cat manner, when he came in search of cyanogas.

Next day was bright enough to begin with, but by nine o'clock it was already clouding over. Croft was still in bed. As Mrs Heyhoe lit the boiler fire, the smoke from the chimney drifted down on to the drive.

"Look at that!" said Marsham, who was waiting with me for the Major to come round. "Tha's always a bad sign, when smoke come low. Today'll be just the same as yesterday. Do you see."

It was.

Everyone made the most of the next fine day. On the top half of Mautby's Hall the kale seed was threshed out. The Upper Prisoners wheat stack was finished by noon, and di-

rectly after dinner carting on the North Field began. Roger
and Leslie cut barley on Clam Breck, Mark and Vincent on
Limekiln Breck. Behind them the women shocked up, one
company in Patfield's charge and one in mine.

"And see they do it properly, with the knots inside," the
Major admonished me, as he proceeded to demonstrate a
well-set-up shock.

"Tha's all very well," observed Marsham as the Major
drove off. "But you can't keep 'em up to it. They're all
wrong the minute you tek your eyes off 'em."

"Then you can't have shown them properly!" I defended
my sex. And went off to show off my new accomplishment
on Limekiln Breck where, sure enough, practically every
shock needed to be reset. Odd sheaves leaned across the ends
of the shocks, blocking the passage of the wind. Nine times
out of ten the sheaves were too upright, inviting the maxi-
mum damage from rain. Knots in the binder string were
conspicuously visible.

"We shall have the Guv'nor after us if he sees these," I
told them, basely throwing all the onus on the absent Major.
"But it won't take us long to put them right." And led the
way round again.

"Have you seen in the papers that land workers can get
some clothes without coupons?" Mrs Roger Marsham
asked me, as we stopped to remove thistle prickles from
our fingers for the nth time. "I got myself a new pair a'
rubber boots yesterday, but I ain't give up the coupons for
'em yet, in case there was something in it."

"Yes. I saw the paragraph. I have to ring up the War
Agricultural Executive Committee about it. But first I

wanted to know roughly how many extra things you actually need. I don't suppose for one minute they'll give you an extra set of coupons, which would be the most sensible thing. But we might as well ask."

One pair of rubber boots a year, one mackintosh, extra jumpers, two or three pairs of gloves, at least half a dozen pairs of stockings . . . everyone was talking at once. The men joined in. "We always reckon to wear out a pair of socks a fortnight in the ordinary way," Bushell told me.

"The Ministry are still threshing out the details," a voice from the War Agricultural Executive Committee came over the wire. "At the moment all you can do is to apply for a permit for special clothing for certain jobs. You have to buy the clothing yourself, it must remain the farmer's property, and it has to be issued for the individual jobs and then called in again."

"But that's quite useless. And the women definitely need and wear out clothes working in the fields that they'd never want if they stayed at home. They can't dig for victory with nothing on."

"That's how the regulations stand at present. They may produce something a bit more useful later."

It rained again during the night. But the day broke fine and wheat carting started almost at once, a three fork company on Palegate and a two fork on Gunspear. The moment the dew was off, barley cutting began, with Bushell and Peel acting as relief drivers and knife sharpeners, so that the binders could work continuously through the day. Before night Mark and Vincent had cut down the Pightles barley and gone off to the Banwell Three Corner. Roger and Leslie finished the Hart Field, and started in Well Piece. The women shocked up, faultlessly this time, helped by the cowmen, once milking was over. Myhill put down his gun and lent a hand. Everybody worked at fever-pitch.

"You're not getting along half fast enough, though. Neither am I," the Major said next morning. And rang up the O.C. of the regiment stationed at West Malton Hall to arrange for six soldiers to help with the harvest at the Manor and another half dozen at the Hall.

"They'll be here tomorrow morning at eight. And be fetched away at seven. You'll have to sign them off. And we have to provide tea. You can get a permit for that from the Food Office. And did you hear on the wireless last night that agricultural leave can be extended another two weeks if we apply at once?"

"Yes. Bushell and Buxton had been in to see me about it and I've rung up the War Agricultural Executive Committee. They're seeing to it right away."

Meale and Old Tom mowed round the edge of the barley on Crabb's Castle and the top of Banwell Close. Twin oat stacks rose in the middle of the latter and another was started on the glebe. Vincent and Mark finished cutting the Three Corner and moved into the Close. Leslie and Bushell helped the carters with the harvest trailers and the Fordson. The Army pitched sheaves off the wagon and on to the elevators, or acted as stackers under Will Long. A thunderstorm tore up the valley, the rain coming down in torrents over Wendeswell, but practically missing us, only shedding a few drops on the eastern side of the farm and stopping Mark and Vincent for less than half an hour. We reshocked the glebe and Coney Hill barley, which the Major thought it might be safe to cart next day.

"Do them storms only keep on a missin' us, we shall hev that all up afore tomorrow night," prophesied Marsham hopefully.

At first it looked as though we were going to be lucky again. We were, insomuch as we had only two points of rain where the Major had eight. But it all came at once in a heavy shower about noon. And after much plunging of hands in sheaf after sheaf of barley, the latter had to be abandoned and Swains' wheat carted instead. Mark and Vincent were able to keep on cutting on Crabb's Castle. Patfield induced several of the women to come shocking up, although the day was Saturday. The cowmen and Myhill joined in. The sun kept out and a light breeze sprang up.

"Looks as though September'll start fine," said Marsham next day, when he came to feed the foals on the Walnut meadow. "Look at all the gnats flyin' around. Well, we can do with it. We ha' hed our share a' wet t'year. The time has bin when I ha' seen barley lyin' about at Christmas. An' I was beginnin' ter wonder if we worn't in fer somethin' a' the sort agen, I kin tell yer."

At long last it looked as though the weather was settling. Though the morning dews grew heavier and heavier, and it was sometimes ten or even eleven o'clock before it was safe to touch the barleys, the sun was scorching for the rest of the day. Wheat was carted first thing, the stacking companies moving over directly the barley was dry enough. Coney Hill, the Hart Field and Well Piece were finished.

And the Major and I set out stacks on Limekiln Breck and the Pightles.

"This should all go on a ten yard stack," he told me, as we paced out the latter, and jabbed elder branches into the ground at each corner. "But the length doesn't matter. It's the width that's important. Barley never wants to be much more than five yards wide. But with a season like this, and layer in it as you've got here, four-and-a-half is ample."

Bushell and Meale were detached from the stacking companies and sent off thatching, beginning with the North Field wheat. Banwell Three Corner, Limekiln Breck and the Pightles were stacked. Mautby's Hall, the last of the wheat, was carted and stacked against the Silford Hill buildings. Mark and Vincent cut down the barley on Leasepit Breck and Doles. Only the twelve acres on Broomhills remained standing.

"And that will have to be cut loose," the Major said. "There's too much second growth there for it to be safe to tie it."

Crabb's Castle was stacked on two stacks, one beside the kale at the bottom of Silford Hill, the other on the far side of the field, on the end of the track running along the edge of the Upper Common. Meale and Bushell took an hour off from thatching to forgather at the carrot washer, with Paynton's engineer, the Major and me (Mark already having taken it to pieces), as a result of which I rang up the War Agricultural Executive Committee and asked for a priority permit for a piece of plate steel. R. A. Leeds called, Root and Hopkins and Manchester rang up, and A. L. Gardiner and three firms with whom the Manor had never yet done business wrote, to enquire about carrots. Mark and Vincent cut down Broomhills, after which the former dismantled and packed up the power-binder while the latter took the Case and the digger plough up to the glebe. The morning dews grew heavier, and were reinforced with fog which hung damply about often until nearly noon.

Towards the end of the week there was one day when the sun failed to break through at all, and it was late afternoon before any barley could be carted, and then only on Leasepit Breck, because the stack was already three parts up.

"Dang the weather!" exclaimed Marsham. "Do that hed only got out, we might ha' finished harvest this week. I never see sich a job."

Another dull day followed, but without the fog. A light wind blew. By nine o'clock barley carting was started again. At the Hunts-and-Gooch side of Doles a twelve-yard barley stack rose. On the other side of the field, beside the Sandy Loke, the Banwell Close barley was carted down and stacked.

"An' le's hope no evacuees don't bu'n that down this time," said Will Long, as he topped it up.

Sunday was a perfect harvest day. The sun shone from the beginning. A light drying wind blew.

"This barley won't be fit to cart tomorrow, though," the Major said as we walked across Broomhills during the afternoon. "It wants at least another day."

Monday started dull, cold and cloudy. The kale seed from Mautby's Hall, shot out on the barn floor until there should be time to deal with it, was re-sacked, and put on rail. At ten o'clock a two-fork company, complete with ele-

vator and stacking-ladders, went over to Ashacre Hall
where, thanks to a series of downpours in which the Hall
fields had been involved but we had mercifully escaped,
the Major still had over a hundred acres of barley to get up.
By noon the sun was out.

I filled in cropping forms for artificial manure permits,
the latest batch of Ministry of Agriculture and Fisheries
Returns, the forms for supplementary cow rations, and
wrote to all the market salesmen and to the Land Army
about the four land girls the Major had decided would be
wanted at the Manor very soon. I rang up the Sedgeham
W.V.S. to arrange (at her own request) for the re-billeting
of Mrs Gaff's mother-in-law, so that the latter could join
her in Flat II. Leslie started ploughing up the barley stubble
on the glebe. Bushell, Mark, and Paynton's man finished
the carrot washer. Peachey and Cecil horsehoed the turnips
on Ram's Breck. Marsham and I tested all the barley stacks
with a stack iron, to be sure that none were getting 'hot'.

Today Broomhills was carted and stacked. Leslie finished
the glebe and moved into the Hart Field this afternoon.
Vincent is riffling up the Banwell Three Corner. Tomorrow
we are washing carrots. Harvest is over and already the new
year is well under way. At seven the army truck rumbled
into the yard to pick up its freight for the last time.

"Funny to think that the war's been on for just over two
years now," the N.C.O. in charge said cheerfully, as I signed
the chit. "Now we shan't be long. Can't be more than another two to go."

"I shouldn't be too sure a' that," said Peacock as the truck
drove off. "The beans in my garden ha' bin growin' the

wrong way round t'year, with their eyes inwards. Never knew 'em ter grow that way afore, except twice. They did it the year we won the Boer War, an' just afore we won the last. Now they ha' done it agen. An' one on 'em ha' twisted itself into the shape a' a 'V' an' all. What do you make out a' that, miss?"

In the skies, the Atlantic, in Europe, Asia, Africa, before the gates of Leningrad, the battles go on. The appalling sum of human misery mounts up and up. The future is a book closed even to the soothsayers in the Sunday papers, in which may be written pages still more hideous that humanity perforce must read. But one thing the land has taught me. Beans or no beans, one day, this year or another, the war machine will run down, the senseless slaughter stop and men learn to sow the seeds for a saner harvest.

"What about that there carrot list?" asks Marsham, as we finish the labour book. "An' Patfield says there aren't more'n a score a' Stebbings' sacks."

"Here's the list. And I've already written to Stebbings."

"Well, I'll be sayin' goodnight, then, miss. An' get along home to me tea."

THE END